Yungcautnguuq Nunam Qainga Tamarmi

❊

All the Land's Surface is Medicine

Yungcautnguuq Nunam Qainga Tamarmi

❋

All the Land's Surface Is Medicine

EDIBLE AND MEDICINAL PLANTS
OF SOUTHWEST ALASKA

❋

ANN FIENUP-RIORDAN, ALICE REARDEN,
MARIE MEADE, AND KEVIN JERNIGAN

with photographs by
Kevin Jernigan and Jacqueline Cleveland

and plant portraits by
Sharon Birzer and Richard W. Tyler

❋

UNIVERSITY OF ALASKA PRESS
FAIRBANKS

Published by
University of Alaska Press
P.O. Box 756240
Fairbanks, AK 99775-6240

Cover and interior design by Kristina Kachele Design, llc.

Cover images by Jacqueline Cleveland. TOP: Kristen Heakin holding up a freshly harvested dock leaf near Quinhagak, July 2019. BOTTOM: Greens Cleveland harvested for her grandmother in mid-June 2016 near their home in Quinhagak, including (from left to right) sea lovage, wild celery, sour dock tops, and wormwood.

Library of Congress Cataloging-in-Publication Data
Names: Fienup-Riordan, Ann, author. | Rearden, Alice, author. | Meade, Marie, author. | Jernigan, Kevin, author, photographer. | Cleveland, Jacqueline, photographer. | Birzer, Sharon, illustrator. | Tyler, Richard W., 1927-2016, illustrator.
Title: Yungcautnguuq nunam qainga tamarmi = All the land's surface is medicine : edible and medicinal plants of southwest Alaska / Ann Fienup-Riordan, Alice Rearden, Marie Meade and Kevin Jernigan ; with photographs by Kevin Jernigan and Jacqueline Cleveland and plant portraits by Sharon Birzer and Richard W. Tyler.
Other titles: All the land's surface is medicine
Description: Fairbanks, AK : University of Alaska Press, [2020] | Includes bibliographical references and index. | English and Yupik.
Identifiers: LCCN 2020013303 (print) | LCCN 2020013304 (ebook) | ISBN 9781602234222 (paperback) | ISBN 9781602234239 (ebook)
Subjects: LCSH: Yupik Eskimos—Ethnobotany. | Ethnobotany—Alaska, Southwest. | Wild plants, Edible—Alaska, Southwest. | Medicinal plants—Alaska, Southwest. | Traditional ecological knowledge—Alaska, Southwest.
Classification: LCC E99.E7 F478 2020 (print) | LCC E99.E7 (ebook) | DDC 581.6/3097984—dc23
LC record available at https://lccn.loc.gov/2020013303
LC ebook record available at https://lccn.loc.gov/2020013304

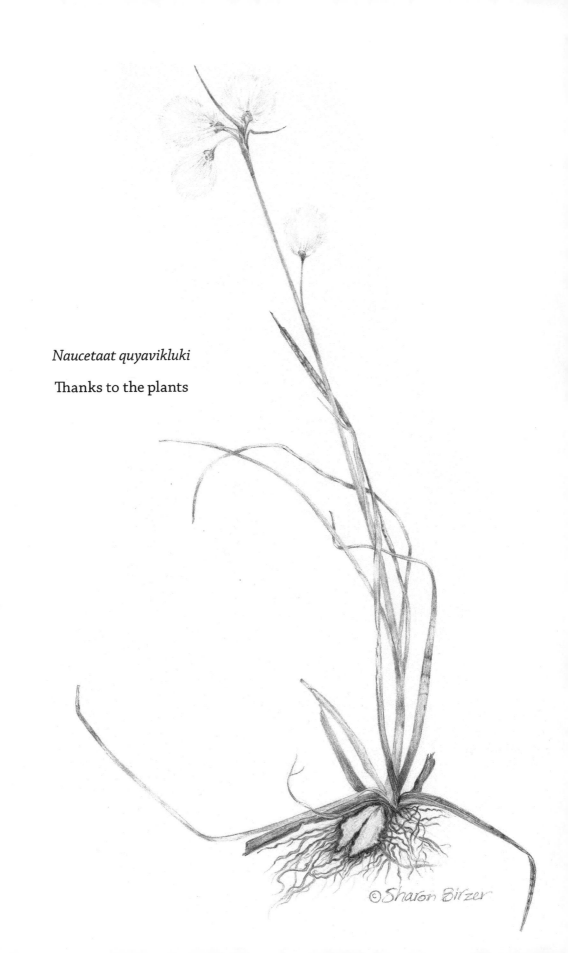

Naucetaat quyavikluki

Thanks to the plants

©Sharon Birzer

Contents

QANEMCIT QULIRAT-LLU
STORIES AND TRADITIONAL TALES

QUYAVIKELPUT
ACKNOWLEDGMENTS

First and foremost, we thank the many elders who contributed to this book over the years. Close to one hundred men and women from all over southwest Alaska shared knowledge of their homeland and the plants that grow there. They are the authors of this book, and these are their stories.

Elders shared this information during gatherings organized by the Calista Elders Council between 2000 and 2014 and Calista Education and Culture Inc. between 2014 and 2019. Funding for these meetings came primarily from the National Science Foundation (NSF), and we remain in their debt. NSF not only gave us the opportunity to work with these particular men and women; NSF also supported a major traditional knowledge project between 2000 and 2005, during which CEC developed the gathering format that we have used over the last fifteen years to document this and many other aspects of Yup'ik life. And an NSF grant between 2016 and 2020 gave us the time and funds we needed to bring together what we had already learned and share it in this publication. This grant also provided an opportunity for three topic-specific gatherings held during February, April, and May of 2019 focused specifically on plants and plant use, during which conversation was guided by large color photographs of different plants as well as plant samples; earlier gatherings had ranged over a variety of topics, including plants. We extend special thanks to our long-time NSF program officer, Anna Kerttula de Echave, for guidance and support throughout this project as well as to our new NSF program officers, Colleen Strawhacker, Roberto Delgado, and Erica Hill, who saw our work through to completion.

Translations of these gatherings, all of which took place in the Yup'ik language, were provided primarily by Alice Rearden and Marie Meade, with additional translation work by David Chanar and Corey Joseph. Alice also contributed her considerable skills to reading and rereading the manuscript to ensure both clarity and accuracy. Most gatherings were organized by CEC cultural director Mark John with the help of other CEC staff members, including Rea Bavilla and Olivia Agnus.

Teamwork has been the hallmark of this book project. Once we had a readable draft, Kevin Jernigan generously provided corrections and advice. He also helped us articulate questions for our final CEC meetings on this topic. There is always more to learn about plants, and Kevin helped us take our work one step further. Along with Kevin, a number of friends and colleagues provided comments on earlier drafts of this manuscript, including anthropologist Dennis Griffin, landscape ecologist Torre Jorgenson, Janet Klein, June McAtee, Dennis Ronsse, and herbalist and author Janice Schofield-Eaton. I also wish to thank the Southcentral Foundation and their health educator Kim Aspelund for welcoming me into their classes and open learning circle regarding Alaska edible and medicinal plants; they not only taught me about plants Yup'ik people use, but about plants available in my own backyard.

Photographs also come from a number of sources. The majority were taken by Kevin Jernigan during his years teaching at the University of Alaska Kuskokwim Campus in Bethel and by Jacqueline Cleveland of Quinhagak during the summers of 2017 and 2019. Other photographs come from the Alaska Botanical Garden, Margaret Samson Beaver, Jeff Foley, J. J. Frost, Ruth Jimmie, June McAtee, the Mabel and Harley McKeague Alaska Inuit Collection at the University of Delaware, James Nicholai, Dennis Ronsse, Talking Circle Media in Anchorage, the US Fish and Wildlife Service, and myself. Once again, my friend and colleague Patrick Jankanish provided a map of the region, and Ian Moore created a detailed map of places mentioned in the text. Seri Tuttle of the Alaska Native Language Center suggested including line drawings of particular plants: Sharon Birzer, with support from the American Society of Botanical Artists, helped us turn this suggestion into a reality, providing beautiful plant portraits of some of our most important plants. Last, the Center for Alaskan Coastal Studies in Homer kindly gave us permission to use color illustrations made by the late Richard Tyler, and Beth Trowbridge, Janet Klein, and Mary Frische worked together in the Center's archives to make this possible.

The orthography used consistently throughout this book is the standard one developed between 1967 and 1972 at the University of Alaska Fairbanks and detailed in works published by the Alaska Native Language Center and others (Reed et al. 1977; Miyaoka and Mather 1979; Jacobson 1995). Plant lists build on the work of Steven Jacobson, published in the *Yup'ik Eskimo Dictionary* (Jacobson 2012). Without their groundbreaking work on the Yup'ik language, not only the plant lists and etymologies but also the detailed transcriptions and translations on which this book is based would have been impossible.

For overseeing the book's publication, we thank the fine staff of the University of Alaska Press, including Nate Bauer, Krista West, Laura Walker, Elizabeth Laska, Dawn Montano, and Amy Simpson, as well as freelance editor Dana Henricks and designer Kristina Kachele.

Finally, we hope our book is as enjoyable for readers to use as it has been for us to put together. It is intended as a guide to the identification and use of plants, but also as an enduring record of what Yup'ik men and women know and value about plants and the roles plants continue to play in Yup'ik lives. We have learned the truth of what Lena Atti once shared: "Our land that grows things is our land. . . . It is so joyous to be there as grasses grow."

One-hundred-year-old Albertina Dull, seated on the left, and her younger sister Lizzie Chimiugak share stories about plants, Toksook Bay, May 2019. ANN FIENUP-RIORDAN

YUPIIT YUUYARAATNEK NALLUNRILNGUUT
YUP'IK TRADITION BEARERS

	Residence	Birthplace	Birth Year
Margaret Andrews/*Kuqaa*	Kotlik		1921
Alma Keyes/*Apaliq*	Kotlik	Pastuli	1922
Martina Aparezuk/*Atangan*	Kotlik	Caniliaq	1932
Angela Hunt/*Yaayuk*	Kotlik	Caniliaq	1940
Mike Andrews Sr./*Angauvik*	Emmonak	Amigtuli	1928
Maryann Andrews/*Tauyaaq*	Emmonak	Qip'ngayak	1933
Joe Phillip/*Panigkaq*	Alakanuk	Alarneq	1923
Barbara Joe/*Arnaucuaq*	Alakanuk	Nunam Iqua	1928
Placid Joseph/*Qavarliaq*	Alakanuk		1933
Lawrence Edmund/*Paugnaralria*			
	Alakanuk	Peguumavik	1934
Denis Shelden/*Kituralria*	Alakanuk	Alakanuk	1944
Anna Pete/*Ac'aralek*	Nunam Iqua	Nunallerpak	1930
Edward Adams	Nunam Iqua	Unguituryaraq	1941
Timothy Myers/*Uparquq*	Pilot Station	Pitkas Point	1926
Marie Myers/*Luqipataaq*	Pilot Station	Cuqartalek	1933
Nick Andrew Sr./*Apirtaq*	Marshall	Iquarmiut	1933
Nastasia Andrew/*Panik*	Marshall	Pilot Station	1935
Francis Charlie/*Acqaq*	Scammon Bay	Anagciq	1941
Neva Rivers/*Aluskaamutaq*	Hooper Bay	Hooper Bay	1920

	Residence	Birthplace	Birth Year
Cecelia Andrews/*Aluk'aq*	Chevak	Qissunaq	1931
Elsie Tommy/*Nanugaq*	Newtok	Kaviarmiut	1922
Susie Angaiak/*Uliggaq*	Tununak	Tununak	1923
Edward Hooper/*Maklak*	Tununak	Tununak	1925
Tommy Hooper/*Cuk'ayar*	Tununak	Tununak	1931
John Walter/*Cungauyar*	Tununak	Cevv'arneq	1939
Helen Walter/*Nasgauq*	Tununak	Tununak	1945
Susie Walter	Tununak		1949
Brentna Chanar/*Papangluar*	Toksook Bay	Cevv'arneq	1912
Phillip Moses/*Nurataaq*	Toksook Bay	Nightmute	1925
Theresa Moses/*Ilanaq*	Toksook Bay	Cevv'arneq	1926
Paul John/*Kangrilnguq*	Toksook Bay	Cevv'arneq	1928
Sophie Agimuk/*Avegyaq*	Toksook Bay	Cevv'arneq	1928
John Alirkar/*Allirkar*	Toksook Bay	Cevv'arneq	1929
Lizzie Chimiugak/*Neng'uryar*	Toksook Bay	Kanerrlulegmiut	1930
Joe Felix/*Amartuq*	Toksook Bay	Manriq	1935
Martina John/*Anguyaluk*	Toksook Bay	Nightmute	1936
Ruth Jimmie/*Angalgaq*	Toksook Bay	Nightmute	1951
Albertina Dull/*Cingyukan*	Nightmute	Qungurmiut	1918
Simeon Agnus/*Unangik*	Nightmute	Nightmute	1930
Anna Agnus/*Avegyaq*	Nightmute	Nightmute	1930
Martina Wasili/*Cuyanguyak*	Chefornak		1924
Maria Erik/*Qamulria*	Chefornak	Cevv'arneq	1926
Jobe Abraham/*Kumak*	Chefornak	Nightmute	1934
Pauline Jimmie/*Kangrilnguq*	Chefornak	Cevv'arneq	1937
Theresa Abraham/*Paniliar*	Chefornak	Cevv'arneq	1941
David Martin/*Negaryaq*	Kipnuk	Cal'itmiut	1914
Frank Andrew/*Miisaq*	Kwigillingok	Kuigilnguq	1917
Lena Atti/*Kayaaq*	Kwigillingok	Qipneq	1925
John Phillip Sr./*Ayagina'ar*	Kongiganak	Anuurarmiut	1925
Carrie Pleasant/*Qak'aq*	Quinhagak	Eek	1930
Martha Mark/*Tartuilnguq*	Quinhagak	Apruka'ar	1934
Alice Mark/*Inaqaq*	Quinhagak	Quinhagak	1934
Joshua Cleveland/*Civialnguq*	Quinhagak	Eek	1937
George Pleasant/*Arnariaq*	Quinhagak	Quinhagak	1938
Annie Cleveland/*Apurin*	Quinhagak	Quinhagak	1940
Pauline Matthew/*Miisaq*	Quinhagak	Quinhagak	1950
Emma White	Quinhagak		1953

	Residence	Birthplace	Birth Year
Katie Jenkins/*Uruvak*	Nunapitchuk	Napakiak	1930
Grace Parks/*Qak'aq*	Nunapitchuk	Cuukvagtuli	1950
Nick Pavilla/*Uruvak*	Atmautluak	Nanvarnarrlagmiut	1940
Marie Alexie/*Akalleq*	Atmautluak	Tagyaraq	1947
Jacob Black/*Nasgauq*	Napakiak	Qaurrayagaq	1940
Ralph Nelson/*Tutmaralria*	Napakiak	Napakiak	1962
Fannie Jacob/*Mayuralria*	Napaskiak	Luumarvik	1932
Nastasia Larson/*Apeng'aq*	Napaskiak	Napaskiak	1940
Martha Evan/*Akiugalria*	Napaskiak	Napaskiak	1945
Marie Andrew/*Akall'aq*	Napaskiak	Napaskiak	1949
Elizabeth Stevens/*Kaukaq*	Napaskiak	Napaskiak	1955
Alexie Nicholai/*Apeng'aq*	Oscarville	Oscarville	1928
Nick Charles Sr./*Ayagina'ar*	Bethel	Nelson Island	1912
Peter Jacobs Sr./*Paniguaq*	Bethel	Cuukvagtuli	1923
Matthew Bean	Bethel	Mountain Village	1932
Olinka George	Akiachak		1932
James Guy/*Qugg'aq*	Kwethluk		1927
Annie Andrew/*Qakuss'aq*	Kwethluk	Napamiut	1933
John Andrew/*Alegyuk*	Kwethluk	Eek Mountains	1945
Wassillie B. Evan/*Misngalria*	Akiak	Napaskiak	1930
Elizabeth Andrew/*Skavv'aq*	Akiak	Tuluksak	1936
Annie Jackson	Akiak		1949
Peter Gilila/*Naparyaq*	Akiak	Akula	1955
Joshua Phillip/*Maqista*	Tuluksak	Akiachak	1909
Mary Napoka	Tuluksak		1916
Bob Aloysius/*Elliksuuyar*	Upper Kalskag	Iinruq	1935
Teresa Alexie	Kalskag		1940
Peter Black/*Nanirgun*	Wasilla	Hooper Bay	1940
Raphael Jimmy/*Angagaq*	Anchorage	Kuiggarpak	1924
David Chanar/*Cingurruk*	Anchorage	Umkumiut	1946
Marie Meade/*Arnaq*	Anchorage	Nunapitchuk	1947
Ann Riordan/*Ellaq'am Arnaan*	Anchorage	Virginia	1948
Mark John/*Miisaq*	Anchorage	Nightmute	1954
Alice Rearden/*Cucuaq*	Anchorage	Napakiak	1976

KALIKAM AYAGNERA
INTRODUCTION

※

A HALF-CENTURY AGO, anthropologists Margaret Lantis (1959) and Wendell Oswalt (1957) wrote that Yup'ik use and understanding of plants were neglected topics and poorly understood by non-Natives. This is thankfully no longer the case. A number of detailed accounts have been written about plant use in particular parts of southwest Alaska: Thomas and Lynn Price Ager's (1980) study of plant use on Nelson Island, Wendell Oswalt's (1957) work in Napaskiak, Margaret Lantis's (1959) study of folk medicine and hygiene in lower Kuskokwim and Nunivak and Nelson Island communities, and Dennis Griffin's (2001, 2008, 2009) and the Nuniwarmiut Taqnelluit (Elders of Nunivak Island) (2018) documentation of plant use on Nunivak Island. A number of excellent studies of Native plant use have also been published in recent decades for other parts of Alaska and the Russian Far East, including books by Lyudmila Ainana and Igor Zagrebin (2014) on Siberian Yupik plant use, Ann Garibaldi (1999) on Alaska Native medicinal flora, Beverly Gray (2011) on wild food and medicine plants of the north, Anore Jones (2010) on Iñupiaq plant use, Wayne Robuck (1985) on plant use in southeast Alaska, Priscilla Russell (2017) on Alutiiq plantlore, Suanne Unger et al. (2014) on traditional foods—including many plants—of the Aleutian and Pribilof Islands, and Janice Schofield's (1989) now classic *Discovering Wild Plants*.

Kristen Heakin holding up a freshly harvested dock leaf near Quinhagak, July 2019. JACQUELINE CLEVELAND

In southwest Alaska, ethnobotanist Kevin Jernigan also worked with a team of elders and Yup'ik language experts to produce an excellent ethnobotany of plants and plant use in the Yukon-Kuskokwim region (Jernigan et al. 2015). Meeting together during four elder councils, men and women from thirteen villages shared 120 plant names corresponding to 73 plant species as well as three species of fungi, two lichens, and one alga. In the book's introduction, Jernigan notes that so much variation exists in Yup'ik plant names and uses that a comprehensive treatment is nearly impossible. The book is therefore intended as an introduction to Central Yup'ik ethnobotany. These caveats notwithstanding, Jernigan and his colleagues have created much more than an "introduction" but rather an invaluable resource without which our present efforts would not have been possible.

This book builds on the work that has gone before it. It is an attempt to bring together information elders have shared about edible and medicinal plants—*naucetaat, nautulit,* or *naumatulit*—and plant use during topic-specific gatherings and field trips sponsored by the Calista Elders Council (CEC, now Calista Education and Culture) over the last twenty years. Beginning in 2000, CEC received support from the National Science Foundation and other federal and state organizations to work with elders and youth in different parts of southwest Alaska to document their history and oral traditions—first in the Canineq (lower Kuskokwim coastal) area, followed by work on Nelson Island, the lower Yukon, Quinhagak, the Akulmiut villages, Kuskokwim communities near Bethel, and most recently, Yukon River communities between Russian Mission and Mountain Village. For unpublished material cited in this book, the meeting date and transcript page number follow the speaker's name in parentheses; at the first mention of each speaker, their home community is also given, e.g., Elsie Tommy (August 2012:29) of Newtok.[1]

More than ninety elders shared stories of their early years traveling on the land, harvesting fish, sea mammals, large and small land animals, and birds. They also spoke eloquently about time spent gathering and storing plants and plant material during snow-free months, including gathering greens during spring, picking berries each summer, collecting wood, harvesting tubers from the caches of tundra voles, and gathering a variety of grasses, which they used for everything from boot insulation to house construction. Sitting on a Nelson Island beach backed by *taperrnat* (rye grasses) and speaking quietly in Yup'ik, Paul John (July 2007:247) of Toksook Bay shared the following:

> With an example, our ancestors urged them to work hard to gather [grass] before there was snow on the ground, the grass that would be twined together, grass for boot insoles, ones they would use for everything, before Ellam Yua [the Person of the Universe] covered the land with his large hand, *Ellam Yuan unalvallraminek nuna pategpailgaku,* as they say.
>
> Snow covers these grasses that grow. They used an example [of Ellam Yua's hands] to encourage their efforts at gathering [grass] before the snow covered them, that [they should gather grass] before Ellam Yua covered the land with his large hand.

Some of what elders described during CEC gatherings was used to create the Yup'ik science exhibition, *Yuungnaqpiallerput/The Way We Genuinely Live,* in 2007. The book that accompanied the exhibition includes detailed sections on both wood and grass—plant materials critical to life in the not-so-

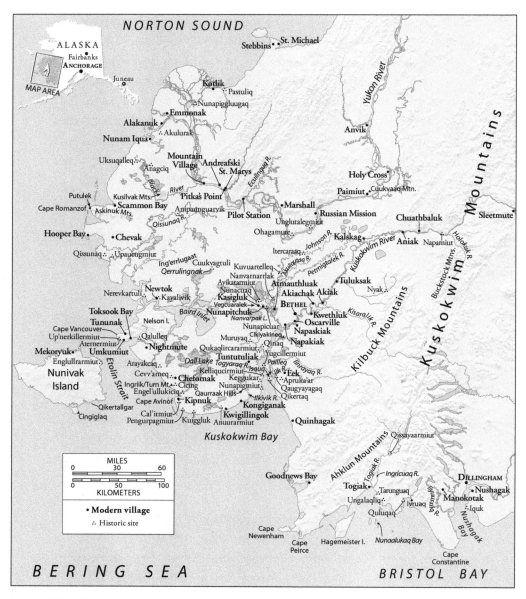

Yukon-Kuskokwim delta, 2020. PATRICK JANKANISH

distant past—as well as the tools used to gather food from the land (Fienup-Riordan 2007). Information on edible and medicinal plants, however, was not included in that book or in any subsequent CEC publication. What follows is our attempt to fill that gap—to share with readers the insights about plants and plant use that elders have shared with us.

As noted, what follows builds on but by no means replaces work by others—especially Jernigan and those who worked with him. Jernigan's work

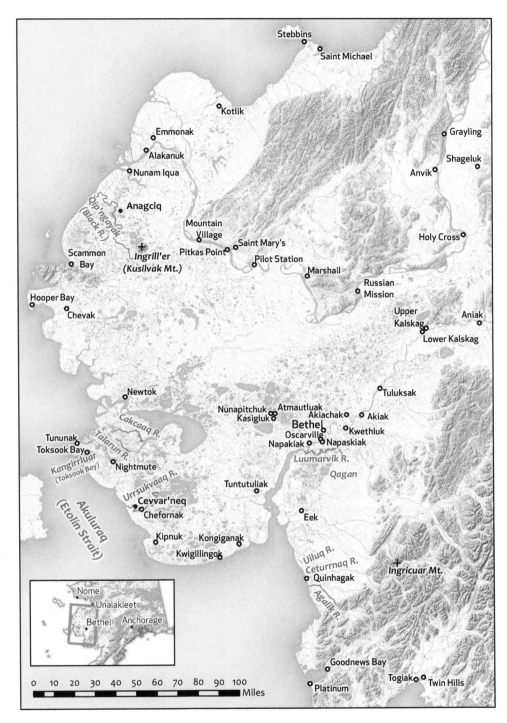

Place names mentioned in the text. IAN MOORE

is much more comprehensive, naming and describing many more species of plants than we describe below. We view what we are sharing—especially the bilingual passages and stories—as enriching our understanding of Yup'ik botanical knowledge and traditions. If the years have taught us nothing else, they have shown us that no single book can hold all of Yup'ik wisdom on any given topic. We share what we have learned in the hopes that others will carry it forward and pass it on.

Winnie Shelden gathering salmonberries at the mouth of the Yukon, August 16, 2011. ANN FIENUP-RIORDAN

An edible mushroom (likely a bolete) growing in a patch of crowberries and Labrador tea glows in the summer light, overlooking the village of Tununak on Nelson Island, August 2018.
RUTH JIMMIE

A field of fireweed in bloom. JACQUELINE CLEVELAND

The Askinuk Mountains, with the Kun River in the foreground. JUNE MCATEE

Lakes to the south of Nelson Island, 2003. JEFF FOLEY

Andrew Beaver harvesting *all-ngiguat* (marsh marigolds) near Kwigillingok.
MARGARET SAMSON BEAVER

➤ Ruth Jimmie and friends gathering wild celery at Umkumiut Culture Camp, June 2016.
TALKING CIRCLE MEDIA

Makirayaraq
Gathering from the land

Southwest Alaska—an area roughly the size of the state of New York—covers a range of Subarctic environments. The delta lowlands between the Yukon and Kuskokwim Rivers are riddled with ponds, lakes, and streams, creating an environment consisting of as much water as land. The Bering Sea coast is dominated by open tundra, with willows growing up along riv-

ers and sloughs. Some shores are rocky, as on Nelson Island, but the coast-
line along both the lower Kuskokwim and lower Yukon Rivers is primarily
sand and mud. As one moves upstream, scattered stands of spruce, willow,
and alder border riverbanks, with tundra and marshland farther inland. The
middle Kuskokwim and Yukon Rivers are dominated by spruce forests as
well as open tundra in the higher country to the north and south.

Mecuqelugaat (sea lovage) growing along the coast near Quinhagak. JACQUELINE CLEVELAND

Though the landscape is vast and imposing, Yup'ik residents have an intimate knowledge of and respect for the plants that grow there. Lena Atti (February 2006:184) of Kwigillingok explained:

> Our land that grows things is our land. We use the things that grow. . . .
> Those things they used. Those places were joyful, including fall camps;
> when we'd first arrive there, it was so joyous to be there as grasses
> grew. . . . And places had names. Our fathers who hunted constantly
> revealed the names. . . . They would mention the name of the place where
> they harvested that particular resource.
>
> We knew the things that were grown that we gathered, even out in
> Kwigillingok. And women knew where they were located, and they knew
> the sloughs down on the coast where they went to gather *taperrnat* [rye
> grasses]. They would mention that they gathered from the shores of that
> particular river. . . . They had names.

Greens Jackie Cleveland harvested for her grandmother in mid-June 2016 near their home in Quinhagak, including (from left to right) sea lovage, wild celery, sour dock tops, and wormwood. JACQUELINE CLEVELAND

The trails they used to gather greens were deep also. When we were eager to get home and on our way, we'd use those trails, and when we used those trails and tried not to divert from them, we'd arrive at the village.

Like Paul John, Neva Rivers (May 2004:300) of Hooper Bay described how the land was like an open hand holding food that could be harvested during summer:

Their hands are open in summer, thus revealing everything. So anything that could be had, could be taken.

Everything is displayed. They want someone to get as much as he can. Berries and plants that will be made into *akutaq* [lit., "a mixture," including berries, fat, boned fish, and other ingredients] are revealed; everywhere they go, there is food to catch.

The wet tundra environment of southwest Alaska, as much water as land. J. J. FROST

So following that, if it goes toward winter, and [Ellam Yua] clenched his hands, there will be no more to catch.

The hand is an example, before it folds that big hand, to get as much as they can so they will not run out during winter.

One who keeps collecting will not go hungry. But one who does not collect with all his might will go hungry. He will not have anything. He will not have any grass. They wanted them to collect those while they were available.

Many declared that the land's surface is full of food. Grace Parks (June 2016:222) of Nunapitchuk noted: "*Nunam qainga neqnguuq* [There is food all over the land]." Mary Napoka (May 1989:33) of Tuluksak recalled: "All things are food on top of the land. They say the surface of the land is filled with food, and the tundra is food. They say when those people of the past used to go hungry, they ate everything." Nick Andrew (February 2019:24) of Marshall said simply: "For people who aren't idle, there is a lot of food in the wilderness."

John Andrew (April 2019:25) of Kwethluk remembered gathering plants and giving them away:

Since I grew up in the wilderness with three sisters, sometimes when I wasn't hunting birds or animals on the land, I would join them, and our mother would have us search for plants in a group. . . .

And when the water was deep, their tops peeked out a little. Those who didn't feel cold would pull them out. . . . When they arrived home, they would divide them among all the people who they had gone with, thinking of their parents and grandparents.

John's (April 2019:26, 29) father had admonished him that if he was hungry, he could eat any plants that birds ate: "I used to be in awe, in the wilderness; they'd tell us that everything that was just growing that a person can eat is food."

John Andrew (April 2019:44) also noted that a variety of foods made one strong: "They never told me that these [plants] were medicine. But they used to say that by not eating just one type of food from the wilderness and adding other things, including plants, to them when we eat, our bodies will be strong. We should not just eat things that have meat, including fish from rivers, but we should also use plants when they grow." Theresa Abraham (February 2019:15) of Chefornak gave a coastal example:

That person from our village mainly ate those plants. . . . She would always travel to the wilderness, even as an old woman, and it seems she never got sick. . . . And she would be gone all day, walking. She'd appear from behind our village when night was about to come holding a small grass bag filled with salmonberries, too. Their mother seemed like she was naturally from the wilderness. She gathered food.

Joshua Cleveland (January 2011:282) of Quinhagak emphasized the importance to this day of keeping the land clean so that plants can continue to be harvested:

Those of us on the [village] council, when the village wants to use the land [around the village], these women tell us not to disturb the places where they gather greens and bird eggs; they tell us to watch out for them when these non-Native people arrive wanting to carry out [construction projects].

They want the places where you gather greens to stay in their original state. Since they gather things from places close to the village, and they even leave on foot, we especially try to make sure that trash doesn't scatter in places where they gather food, or that they don't dump contaminants around those places. . . .

And there's a fence around that [dump], and it isn't placed close to the village. Following the women's request, not wanting it to be too close to the village, keeping in mind the place where they gather greens, thinking that those places where they usually gather things will disappear, we watch out for that when we make plans for the development of our village. It's good as their gathering places are still there because they use them year after year.

David Martin (May 2004:17) of Kipnuk emphasized plants as a source of wealth:

That's how they are in my observation. They gather the roots of plants that grow on our land. And those people gathered as much as they could. When I observed those past people, since they didn't have what the white people called purses, they only had *issraksuaraat* [small grass carrying bags]. They would store the fish they caught during summer in grass containers. Since the only source of wealth that they had was plants, the women would gather all different varieties of those during summer when they grew.

Wassillie B. Evan (March 2004:556) of Akiak also spoke of plants as wealth when he told the story of two who attended a dance festival among *ircenrraat* (other-than-human persons). When they arrived, their hosts advised them to choose as gifts things that others didn't want to take, including grasses, *pellukutat* (coltsfoot leaves) and *qanganaruat* (wormwood). When they reached home, they threw the bags that they had filled with plants into their storage cache. After some time, when her husband went to check his cache, he saw that the skin container was stuffed full. When he opened it, he saw that the *pellukutat* that he had placed inside had become bearded-seal skins. And the *qanganaruat* (lit., "pretend squirrels," said by some to be the squirrels of *ircenrraat*) had become squirrel skins: "The plants that they had placed inside in their village had transformed into those things." Plants can indeed become a source of wealth.

In part because plants are so highly valued, their harvest is celebrated. When a young girl picks her first berries, her family will often give gifts in her honor, either immediately or at a dance festival during the following winter. The berries themselves will often be given to a village elder or close relative, ensuring the child will continue to harvest in abundance.

During critical periods of one's life—the death of a spouse or loved one, a miscarriage, a girl's first menstruation—*eyagyarat* (abstinence rules) are followed in many communities to this day. *Eyagyarat* codify the special care people must exercise during life's transformations, and they often involve refraining from harvesting activities, including gathering greens, grass, and berries for a specified period of time. Following her first menstruation, a girl must not expose herself to the world generally but must keep herself covered both indoors and when exiting her home. When she does go out, she must carry on her person a small sack of ashes made from burned plants. Following a period of restricted activity, she can travel again, but only after sprinkling ashes across her path. All of these restrictions are enacted to protect those in vulnerable positions from the universe, which is said to be aware of their situation. To ignore these restrictions endangers both the person immediately involved as well as the larger community. As testament to the sentient world in which she lives, Marie Andrew (March 2017:88) of Napaskiak shared her grandmother's advice: "She used to say, 'Everything, including plants, trees, and those residing in water, has a spirit. That is why we humans observe the *eyagyarat*.' When I had my first menses, she told me not to pull plants from the ground [to gather them]." Martha Evan, also from Napaskiak, explained that this was because living things, including plants, can harbor ill feelings and retaliate.

Yupiit Kass'at-llu qaillun naucetaat umyuartequtellrit
How Yup'ik people and others think about plants

Before continuing with discussions of particular species, comments on how Yup'ik plantlore is like and unlike Western botanical knowledge may prove helpful. First, Yup'ik people use the words *naucetaat, naumrruyiit,* or *naungrruyiit* for plants and flowers of all kinds—including grasses, shrubs and trees, edible greens, and medicinal plants. These Yup'ik terms for plants also include fungi, lichens, and algae, while Western botany does not, strictly speaking, classify these three groups as plants.[2]

Also like Western plantlore, Yup'ik people often have different names for the same plant species.[3] For example Yup'ik terms for *Artemisia tilesii* include both *caiggluut* and *qanganaruat,* while common Western names include both wormwood and stinkweed. People in different villages know similar plants by different names. Angela Hunt (December 2012:74) of Kotlik noted: "In this Yukon River and farther downriver, they have different [names]. When I was growing up, we called those that grow yellow flowers on top during the summer *irunguat* [marsh marigolds]." Anna Pete of Nunam Iqua added: "We call them *allngiguat* [marsh marigolds] down on the coast." In some cases the same Yup'ik name is used in different areas for two distinct plant species, and we have added clarification in our discussion wherever possible.

The pages that follow document the most widely used and well-known plants, omitting several species that elders did not discuss. However, we also include a small number of plants that elders described in detail but for which we have been unable to determine the English name. We do so hoping that in the future others can use these descriptions to identify these "mystery plants." Sadly, plant knowledge in southwest Alaska is dwindling (Ager and Ager 1980:39; Nuniwarmiut Taqnelluit 2018:182). John Andrew (April 2019:34) remarked: "They say many things on the land can be medicine, but we don't know what most of them are. And those younger than us really don't know them because they believe in Western [medicines]."

We have roughly organized information on plants and plant use by seasonality, with the most popular and well-known plants described first, followed by discussions of lesser-known plants. Thus, we start with edible plants gathered during spring, moving on to plants gathered in early summer. Many plants, such as *tayarut* (mare's tail) and some plant tubers, are harvested both in fall and spring, and these—along with several lesser-known edible plants—conclude our discussion. The section on edible plants is followed by a long discussion of medicinal plants, many gathered later in the summer and dried for use all year round, as well as a brief but important section on poisonous plants to be avoided. This is followed by discussions of berry picking, beginning in July and continuing through August and early September, followed by hunting for plant tubers in the caches of tundra voles just before freeze-up in the fall.

Many of the men and women we spoke with noted that people in different villages ate and used many but not all of the same plants. Martina John (November 2011:325) of Toksook Bay said: "[We gather plants] based on our knowledge of them. And when we travel somewhere, we see that plants that we don't usually gather are gathered by those people. We weren't aware that they were edible. That's how it is since all villages are different." Conversations were particularly interesting when an elder described a well-known plant from his hometown that others did not recognize. A memorable example was when Bob Aloysius (October 2010:127) of Kalskag talked at length about *elagat* (alpine sweet vetch), commonly known as "Eskimo potatoes." *Elagat* grow along rocky stretches of the Yukon and upper Kuskokwim Rivers but are not found along the muddy coasts of the Canineq area. John Phillip of Kongiganak was interested in learning about this plant, which Bob described as delicious.

Names for particular plants often tell us a great deal about them, and we have included etymologies whenever possible. When speaking of plants generally, a Yup'ik speaker can refer to *acilquq* or *nemernaq* (root), *avayaq* or *cuyaqsuk* (branch), *cuya* (leaf), and *epulquq* (stem). Flowers and blossoms (as well as plants generally) are referred to as *naucetaat*; a flower petal as *caqel-ngataruaq* (lit., "imitation butterfly") and the sepals as *naucetaam caqelnga-taruarita nayumiqassuutait* (lit., "supports of the flower's petals") (Jacobson 2012:185). *Qulirat* (traditional tales) involving visits to the underground world of *ircenrraat* sometimes make reference to *nunam taqra* (the earth's vein), which cannot be cut. Human visitors who succeed in breaking *nunam taqra* are said to have been immediately sent back to the aboveground world where humans dwell.

Unlike humans, plants do not die but rather wither. Marie Meade (April 2015:358) from Nunapitchuk commented: "They do not say that [fish or plants] *tuquluteng* [die]. The fish *nalaluteng* [die, wither] or the plants that grow on land also *nalaluteng*. They speak of *enrilriit* [those that become fleshy] when they speak of plants that are grown. People are probably the only ones who *tuquluteng* when their breath leaves them."

One important feature of Yup'ik botany is the distinction people make between male and female parts of a plant. Raphael Jimmy (January 2013:353), originally from Nunam Iqua, declared: "Everything on the land, even if it was a plant, they knew whether it was male or female. They would say that this particular plant was a female. They knew what they were. And these grasses have males [and females]." Generally, the flowering and seed-bearing parts of a plant are considered *angucaluut* (males), while the leaves are considered *arnacaluut* (females). Species where this is the case include *caiggluut* (wormwood), *ayuq* (Labrador tea), and *nasqupaguat* (nutty saw-wort). In some cases one plant is male and another female, each with its own name. For example, some consider *quagcit* (sour dock) to be the female mate of *nakaaret* (wild rhubarb), and *ikiituk* (wild celery) as the male of *mecuqelugaq* (sea lovage). Wendell Oswalt (1957:19) notes that lichens, horsetails, wood ferns, and mosses are not classified by sex.

Anna Roberts harvesting *kapuukaraat* with a hook and staff, May 9, 2019. JACQUELINE CLEVELAND

NUNAM QAINGANI
NERTUKNGAIT
INRUKTUKNGAIT-LLU

EDIBLE AND
MEDICINAL PLANTS
OF SOUTHWEST ALASKA

NAUCETAAT NERTUKNGAIT
EDIBLE PLANTS

∗

Makiralriit up'nerkami
Those gathering greens in spring

Makiralriit [Those gathering plants and greens] are like this. Those who go
to gather greens and plants go in large groups. They especially go during the fall
and spring. . . . They can no longer stay put when the weather is good.
—Frank Andrew (February 2003:841), Kwigillingok

Spring is a time of warming weather and new growth. As ice melts and snow
pulls back, women and men begin to gather *cuassaat* (wild, edible greens)
all over southwest Alaska. Wassillie B. Evan (July 2000:26) recalled: "There
are plants that grow in the spring that people ate like *allngiguat* [marsh
marigolds], *iitaat* [tall cottongrass], *angukat* [wild rhubarb], *quagcit* [sour
dock], and *kapuukaraat* [buttercups], and these plants kept the people of the
past from getting sick." Nick Charles (December 1985:138), originally from
Nelson Island, said simply: "All the plants that are just sprouting that we
eat from the land will help our bodies." Neva Rivers (May 2004:73) noted:
"When it's time to gather *aatunat* [sour dock], we add fish roe to them, and
they're good eating. . . . Those *kapuukaraat* [are harvested] during spring, and
in the middle of June. And when the *nasqupaguat* [nutty saw-wort] have just
started to grow, they add them to fish."

Martha Mark (January 2011:238) of Quinhagak noted the early spring greens at the mouth of the Kuskokwim:

> We [gather greens and eggs] from around our village. And there are many *it'garalget* [beach greens] down along our shore. They extend down the coast. They obtain *it'garalget* from there.
>
> And *tayarut* [mare's tail plants] are in small lakes in the fall and in summer. They are good when picked while they are green, cooking them and making them into *akutaq*.

Maria Erik (March 2008:223) of Chefornak remembered:

> They would gather and store *allngiguat* [marsh marigolds] in June, and the *quagcit* [sour dock] are short at that time. And down on the coast, downriver from our village, are *nasqupaguat* [nutty saw-wort] also. Some time ago I pulled a few, picking some for Iraluq to eat. He evidently really enjoys eating *nasqupaguat*. Then when I arrived, I brought them up to shore. The person who I gave them to cooked them. I didn't find them tasty in the past. When I ate a little, I started to find them tasty. I regretted having given them away. [*laughs*]

David Martin (May 2004:48), also from the coast, noted that *kapuukaraat* (buttercups) were the first plants people gathered: "They gathered those first, with winter in mind, [gathering] starting in spring. And while we were [camped on the coast], we would gather *kapuukaraat* during springtime, but they never stored those for eating later, since there was nowhere to store them, but they immediately cooked them and ate them. *Kapuukaraat* are the first plants that we gather."

Lizzie Chimiugak (January 2007:231) of Toksook Bay remembered the long distances women traveled to harvest *kapuukaraat*:

> Women in spring, when they used to try to get food from the land, when our ancestors did not have trails, they used to get *kapuukaraat* from up there [toward Nightmute] by walking from Up'nerkillermiut.
>
> And somewhere up there, upriver from Qalulleq, they would [close] their *issratet* [twined grass carrying bags] and go back home [to Up'nerkillermiut] because they were so packed full of *kapuukaraat*. They would pack them in and tie them securely.
>
> How daunting those people were at that time, the former inhabitants

of Up'nerkillermiut. When plants were ready, including *kapuukaraat*, even while there was ice, they used to appear.

Lizzie Chimiugak (January 2007:231) also recalled harvesting the coastal plant *ariqait* or *ariqat* (English name unknown) in early spring:

> And when they ran short of food long ago, from up above them, they dug *ariqait* by chipping away the ice. Just to let their children be occupied [using them as pacifiers] they let them eat *ariqait*. My mother told about that because she experienced it. . . .
> *Nunam qainga neqnguyaaquq* [The top of the land is indeed full of food]. . . . And their women did not sit around idle. They were so daunting when she talked about them.

Although most edible greens are not considered medicinal plants, their health benefits in spring were appreciated. During our discussion of a variety of spring plants, including *mecuqelugaat* (sea lovage), *ikiituut* (wild celery), and *quagcit* (sour dock), Albertina Dull (November 2007:307) of Nightmute exclaimed: "People are actually consuming medicine, really consuming medicine. When [plants] grow, people constantly consume medicines. They eat them and only stop consuming them when they harden."

Kapuukaraat
Buttercup plants

Many recalled harvesting newly sprouted *kapuukaraat* (buttercups, *Ranunculus pallasii*) in spring. Theresa Abraham (November 2007:321–23) described wading into shallow ponds and using the tip of her walking stick to cut and harvest *kapuukaraat*. The name *kapuukaraat* derives from *kapur-* (to poke or stab) and probably refers to this harvest method.

> They use a wooden tool with a [curved] tip. It's pretty long, and they said they used it as an *ayaruq* [walking stick]. They said long ago, women always used the walking stick every time they went out to the wilderness. And when we were about to go on lakes, they always had us use a walking stick. . . . The tip along the end was used to check the area where we were about to walk. We always [jabbed] the place where we were about to go down [into the lake] in spring.

Kapuukaraat (buttercup plants), mature and much too late in the season to harvest. KEVIN JERNIGAN

Then along our walking sticks, the tip that looks [curved], we used it to gather *kapuukaraat*, using it to cut it down. Then we picked those *kapuukaraat* and placed them in our containers. . . .

If the ice won't break, we go down. . . . After sticking our walking stick inside the water and searching for them, using the top of the tool, we cut them under the water along their base, and then those appear on the surface.

Martha Mark (January 2011:239) also described gathering *kapuukaraat* in coastal ponds near Quinhagak: "When spring comes, *kapuukaraat* are first to be available in lakes. . . . We just scoop [those buttercup plants] out in lakes along mosses. We wade in the water with walking sticks. Some [lakes] don't have any, and some have a lot." Pauline Matthew (January 2011:240), also of Quinhagak, noted that *kapuukaraat* are available during the end of April or the beginning of May in lakes behind Quinhagak when the bottom of a lake is still frozen: "We get those *kapuukaraat* from lakes before they are ice-free, alongside those mosses. They are obvious around mosses as their ends are showing. Then using our walking sticks, we [reach out] and bring the mosses that have those *kapuukaraat* with them up onto land. Then after cleaning those, we bring them home."

Anna Roberts harvesting *kapuukaraat* with a hook and staff, May 9, 2019. JACQUELINE CLEVELAND

Martha and Pauline agreed that later in the season, when *kapuukaraat* begin to mature, they have a bitter taste and are no longer good to eat. Martha commented: "When they grow, they start to sting when eaten, and we stop gathering them." Pauline added that she had heard that buttercups grow twice, but that the second growth are like peppers, and many do not eat them. John Andrew (April 2019:28) noted that some still harvested *kapuukaraat*, even after they flowered: "Our older sister and my mother said they used them like pepper. . . . They put them in soups for flavoring." Cecelia Andrews (April 2019:28) from Chevak agreed: "They didn't stop gathering them right away, they'd add them to soup, even though they became a little spicy. They were very delicious." Theresa Abraham (April 2019:47) remarked that *kapuukaraat* can also be gathered in the fall: "We heard that they don't become inedible. . . . But in the fall, close to the time when lakes are going to freeze, they become good to gather again, they become like they had been in the spring."

Anna Roberts cleaning buttercup plants as she harvests them. JACQUELINE CLEVELAND

John Andrew (April 2019:42) noted the use of *kapuukaraat* in times of scarcity:

> They say when Western goods were not available, they worked hard to pick them, before fish and birds became abundant.
> Some years they are gone in the places where they usually grow. They would say they had moved to another place, and even to another lake.

Pauline Matthew (January 2011:242) described the many uses of this tasty plant: "We use *kapuukaraat* when we make bird soup. Or when we eat half-dried and boiled fish, we just place them inside [the broth] as they are and eat them. Or we use them in *qayussaaget* [broth soups]." *Kapuukaraat* are not usually eaten raw, and in fact cooking is important, as it destroys toxic alkaloids present in the raw plant; at the same time, one must be careful not

Anna Roberts and her daughter Dana with the rewards of a good day's work. JACQUELINE CLEVELAND

to overcook them. Theresa Abraham (November 2007:323) explained: "They check them when they cook them. When they are overcooked, they tend to disintegrate like those *negaasget* [silverweed tubers] or *utngungssaraat* [grass seeds]. . . . But when checking them while cooking them, they cook them just right [not broken up]. Then they put them in soup or eat them on their own after cooking them. They are delicious." Grace Parks (June 2016:226) noted that *qaqacuqunat* (another term for buttercups, possibly from *qaqaq*, "red-throated loon") are also added to soup in the Akulmiut area: "They are very delicious when added to broad whitefish and to king salmon."

Buttercups. SHARON BIRZER

It'garalget
Beach greens

Another well-known plant available in early summer is *it'garalget* (beach greens, sea chickweed, *Honckenya peploides*, lit., "ones having little *it'gat* [feet]"). Other Yup'ik names for beach greens include *itegaraq* (from *itegaq*, another word for foot), *qelquayak* (from *qelquaq*, "rockweed or kelp"), and *tukulleggaq* (from *tukullek*, the Cup'ig (Nunivak) word for foot). Martha Mark (January 2011:239) recalled: "When the time to harvest *kapuukaraat* passes, they gather *it'garalget*." Pauline Matthew (January 2011:241) accurately identified the sandy environment where beach greens grow: "*It'garalget* grow down along the ocean shore all the way down to Platinum and Goodnews Bay on the sand. And they also grow extending up the coast on the sand." Martha added: "They don't grow at Uiluq [River] because it's mud." Many note that *it'garalget* are delicious when eaten either raw, dipped in seal oil, or cooked and added to soups or *akutaq*.

When beach greens start to grow flowers and harden, they are no longer harvested. Martha noted: "They say they start to light up, *kenurrangluteng*, when they start to turn yellow." Ruth Jimmie (November 2007:314) of Toksook Bay recalled the same saying on Nelson Island: "When they form flowers, they say *kenengluteng* [they light up]."

It'garalget (beach greens) growing along the coast near Quinhagak. JACQUELINE CLEVELAND

Molly Alexie and her daughter, Nicolette, harvesting beach greens near Quinhagak, June 2, 2019.
JACQUELINE CLEVELAND

Like *kapuukaraat*, beach greens are said to grow a second time. Martha noted: "The first ones that grow get hard, and they grow yellow things on them. Then in July, others grow along their branches. Those are good and soft." Martina John (November 2007:314) observed the same thing on the rocky beaches of Nelson Island: "When the land thaws, those *it'garalget* grow right away. And they grow over and over again. After wilting, they grow again." Ruth Jimmie (April 2019:50) added: "These [*it'garalget*] grow all summer. In the fall, when they grow, they are very big, and sometimes they grow small balls. Mary Matthias [from Nightmute] gathers them when they are like that, and she says they are more juicy."

> Nicolette Alexie holds a handful of beach greens. JACQUELINE CLEVELAND

▲ Molly cutting off the bitter ends of the beach greens. JACQUELINE CLEVELAND

Allngiguat
Marsh marigolds

Allngiguat (marsh marigolds, *Caltha palustris*, lit., "pretend *allngik* [boot sole patch]," which its leaves resemble) are also known as *irunguat* (lit., "pretend *irut* [legs]") in the Kotlik area, *allmaguat* in Hooper Bay, and *uivlut* on Nunivak Island (from *uive-*, "to circle," because of their round shape). Like *kapuukaraat*, *allngiguat* are harvested from ponds as well as from the muddy shores of sloughs in spring. Theresa Abraham (February 2019:126) noted that people do not harvest the female plant, but only the male plant with buds before they flower. Unlike *kapuukaraat*, however, they do not grow a

Marsh marigolds. SHARON BIRZER

➤ Marsh marigolds with their blossoms still tight, ripe for harvesting, June 6, 2019. JACQUELINE CLEVELAND

▲ Marsh marigolds in bloom. KEVIN JERNIGAN

Jackie Cleveland lightly cooking marsh marigolds at home in Quinhagak. JACQUELINE CLEVELAND

second time, and when yellow flowers appear, people stop gathering them. Pauline Matthew (January 2011:243) explained: "Those *allngiguat* grow along the shores of streams and have small yellow things on them. They gathered those marsh marigolds before they bloomed." Neva Rivers (May 2004:70) said: "When it approaches summer, those *allmaguat* are available, when they've grown a little taller following the grass."

John Andrew (September 2018:140) remembered watching his father harvest them in spring: "He used clothes that wouldn't get wet, even *ivrucik* [waterproof skin boots]. After wading, he would pull some plants out, and their roots looked like water weeds. They say that year after year, those grow in the same place." John (April 2019:28) said that he used to search for them himself in lakes: "Their stems are long in deep places, and short in shallow areas or places that are dry." John (September 2018:140) had seen *allngiguat* growing along warm spots in mountain streams during winter, and he said that mallards used them to survive: "We see those [ducks] flying around. They are a sight to see in the middle of winter."

Fannie Cleveland Moore enjoying marsh marigolds with seal oil. JACQUELINE CLEVELAND

Allngiguat are always cooked before eating to leach out toxic chemicals (Jernigan et al. 2015:65). Martha Mark added that *allngiguat* were good to eat with half-dried and boiled fish, like salmon. And John Andrew (April 2019) said that they are good when added to animals that are fatty, such as fresh sea mammals or birds. Pauline Matthew noted with regret that today few people gather them.

Ulqit
Tuberous spring beauty

Ulqit (tuberous spring beauty, also known as "Eskimo potatoes," *Claytonia tuberosa*) are found growing in the coastal mountains of Nelson Island. Nightmute elder Albertina Dull (May 2019:169) described harvesting them and eating their roots in spring:

Ulqit (tuberous spring beauty). KEVIN JERNIGAN

> Up there [in the mountains], you know when those plants grow in spring, they have small white tops. When we'd go up there right to the summit, those have little stems like *naunrat* [salmonberries], and [their flowering tops] are white. People would dig those up.
> And under them, they are very white and soft, people would eat them. People called those *ulqit*. They are delicious. Their insides are very white. They're small.
> I think they only grow around Up'nerkillermiut, up there above [in the mountains].
> You know, their top parts easily come off, they're blown away, but their buds are [short] like ones that will become *naunrat* [salmonberries]. . . . We knew when we saw them, and we would dig them up right away and eat them.

Albertina's description is a good match for information recorded in the late 1970s by Thomas Ager and Lynn Price Ager (1980:30): "Several children and adults described a plant that grows in the tundra near Tununak which looks quite similar to *Claytonia sarmentosa* (which is not edible) and may be *Claytonia tuberosa*. It has a walnut-size underground corn that is edible. Children said they often eat them raw when out on the tundra."

Tuberous spring beauty, showing its thick edible tuber. KEVIN JERNIGAN

Cetuguat wall' ceturqaaraat
Fiddlehead ferns

Ferns are another spring delicacy. People know them by different names, including *cetuguat* (lit., "pretend *cetuut* [fingernails or claws]"), *cetugpaguat* ("pretend long fingernails"), and *ceturqaaraat* (from *cetur-*, "stretching one's legs"). Jernigan (et al. 2015:134) notes that these names cover two different fern species, *Athyrium filix-femina* (lady fern) and *Dryopteris expansa* (wood fern).

Many gather *cetuguat* just after they emerge as tight, curly balls and before they unfurl. John Andrew (April 2019:62) remarked: "They use them upriver only in spring. They tell them not to gather them when they harden, they say when they had stretched out, *ceturcimariluteng*. That's why they call them *ceturqaaraat*." Pauline Matthew (January 2011:243) said: "We gather those in May, toward the end of the month. It's good when picking them when they are just about to open up. If you pick them before they open up, it's really tiring to clean them." Cleaning ferns requires rubbing away the brown hairs clinging to the new growth. Some say that these hairs can be eaten on their own to help digestion (Jernigan et al. 2015:135).

Fiddlehead ferns are found in a variety of locations. Pauline Matthew (January 2011:244) said that she harvests them in places where the land has caved in, or where the tundra drops off into marshland. Farther upriver, Elizabeth Andrew (April 2008:222) of Akiak said that she finds them in the vicinity of trees and bushes, while Susie Walter (April 2008:222) of Tununak finds ferns among willows. Ruth Jimmie (April 2019:306) recalled picking the tops and playing with them: "We put them here [on our fingertips], pretending they were our nails." All agreed that ferns are good when cooked and added to soup or *akutaq*. Francis Charlie (January 2013:347) of Scammon Bay declared: "*Cetuguat* are extremely delicious in *akutaq*."

In some areas, fern roots are also harvested. Martha Mark (January 2011:245) noted: "The roots of *ceturqaaraat* are *kun'at* [edible roots of spreading wood fern, *Dryopteris expansa*]. The small things are long. You dig underneath [the fern], and then underneath they are very small and white. *Kun'at* taste very sugary and are delicious." George Pleasant (January 2011:245) of Quinhagak added: "They are good cooked. I compare their taste to potatoes. They dig and obtain them before the plant above them grows."

➤ Molly Alexie harvesting fiddleheads near Quinhagak.
JACQUELINE CLEVELAND

Cetuguat
(fiddlehead ferns)
growing near
Quinhagak,
June 1, 2019.
JACQUELINE
CLEVELAND

Mecuqelugaat wall' mercurtulgaat
Sea lovage

Sea lovage (*Ligusticum scoticum*), found near beaches along the shores of the Bering Sea and enjoyed both raw and cooked, is known by a variety of Yup'ik names; some names derive from the base *mecuq* (liquid part of something, sap, or juice) including *mecuqelugaat*, *mecurtulit*, *mecuqellugaq*, and *mecuggluggaq*, while *mercurtulgaat* and *mercurtuliaraat* derive from the verb base *mer-* (to drink). Jernigan (et al. 2015:74) notes that in Hooper Bay and Chevak, sea lovage is also termed *ikiitum arnacalua* (female *ikiituk* [wild celery]) for its resemblance to the larger plant, and Helen Walter of Tununak referred to them as *ikiituyagaat* (small *ikiituut*). Martha Mark (January 2011:248) added that they grow around *taperrnat* (rye grass, *Leymus mollis*) along the ocean shore.

Quinhagak elders noted that *mecuqelugaat* are eaten with half-dried and boiled fish. They can also be made into *akutaq* or enjoyed fresh without being cooked. George Pleasant (January 2011:248) remarked: "*Mecurtulit*. When I

◀ Molly Alexie and her fiddlehead harvest. JACQUELINE CLEVELAND

▼ *Mecuqelugaat* (sea lovage). KEVIN JERNIGAN

Mercuqelugaat (sea lovage) growing along the coast near Quinhagak, June 16, 2018.
JACQUELINE CLEVELAND

see some down along the ocean shore, I cut them and eat them. Eating the ones that have just grown without cooking them is also good." Martina John (November 2007:307) said: "We used to eat *mecurtuliaraat* when we'd play. . . . We would take them along their ends and eat them." Albertina Dull added: "They [gather them] along with their stems; they also pull their stems and place them in seal oil, and they eat those when eating dried foods." Martina agreed: "We ate them without cooking them, adding them to fish." Ruth Jimmie (April 2019:93) concluded: "There are many of these *mercurtuliaraat* in our home area. . . . When we gather these, we put them in seal oil and then when we eat dry fish, we supplement our meal with them. . . . They don't grow just anywhere. And even at Umkumiut they pick them inside that valley. . . . They're delicious."

Elquat epuit
Rockweed

Technically an alga, Jernigan (et al. 2015:147) identifies *elquat epuit* as rockweed or bladderwrack, *Fucus gardneri*, a brown alga growing along rocky shorelines in southwest Alaska. Herring lay their eggs there each spring, and people harvest and eat the eggs and algae together, dipped in seal oil. This salty combination is known as *elquat*, and the rockweed itself as *elquat epuit* (lit., "stem for *elquat* [herring eggs]"), and *elquat* or *qelquat* for short. Mark John (February 2019:24) commented: "In the ocean, too, those they call *elquat*, my grandmother said those can be eaten even if herring haven't laid their eggs on them. They grow on rocks where herring lay their eggs."

Elquat epuit (rockweed). KEVIN JERNIGAN

Elagat wall' qerqat
Alpine sweet vetch

Bob Aloysius shared a detailed account of gathering roots known as *elagat* (alpine sweet vetch, *Hedysarum alpinum*, or "Eskimo potatoes," lit., "things dug from underground") along the middle Kuskokwim and *qerqat* on the lower Yukon:

During spring, ice usually scrapes the shore, and then *elagat* would appear. Some of those *elagat* were about [one inch] wide and were long. We'd dig the ground and remove them and after washing them, we'd cut them to pieces.

We'd mainly make them into *akutaq*. We'd also add wild rhubarb to them, sheefish meat, or a mixture of both. Those were extremely good!

White people call them "sweet potato roots." They're really long. They're about [18 inches] tall, green, with lots of leaves, and their tops, lots of pink flowers. . . .

We get them from on top of gravel beaches because they grow on the gravel. And when the ice scrapes them, they get exposed, and you get them out. . . .

They are like roots. They grow all over underground. Some are extremely long. Some are about the length from here to that corner over there [eight feet].

They are extremely tasty in spring. Those are available after the ice breaks up and leaves. They pry them [from underground]; that's why they are

Alpine sweet vetch and its roots. SHARON BIRZER

Alpine sweet vetch. KEVIN JERNIGAN

called *elagat* [lit., "things dug from underground"]. They dig them out of the ground, cut them in pieces, and store them away. And after picking them, some people put them away after drying them, or they just place them in a plastic bag. They don't dry. They take them out in winter and make them into *akutaq*.

They are very tasty, tasting sugary.

Nick Andrew noted that *elagat* are also found downriver from Marshall, while John Phillip declared they are not found on the muddy shorelines of the lower Kuskokwim coastal area.

As noted above, *elagat* are also known as *qerqat* on the lower Yukon. Raphael Jimmy (January 2013:337) described his grandmother adding the tops of *qerqat* to soups:

When she was cooking she would tell me to do the following, "Go and get the tops of *qerqat*." I would get a great many.

Then after ladling the food into bowls, she would put those inside [the pot] and cook it at a low boil and then she turned it off. Then she added that soup to our bowls. Boy, they're good.

She added the tops of *qerqat*, the purple flowers. You can eat them around August. They are really sweet.

Francis Charlie added that they are also good raw: "When taking them and pulling them, one's hand becomes full and one eats them and they are delicious." Raphael Jimmy (January 2013:338) was warned not to overeat: "In the past our mother told us not to eat too many of those tops of *qerqat* without seal oil, but she said it was okay if we ate a lot if we ate them with seal oil."

Raphael continued: "The tubers of those *qerqat* are underground. In the past, they were Yup'ik potatoes. . . . *Avcelngaat* [voles] always gather those for food." Francis added: "They look like potatoes, but they are long and have many branches. Their skins look like potato skins." Angela Hunt (December 2012:81) described gathering *qerqat* near Kotlik: "They travel all over and they also pry the ground for those close to the coast. Some have many

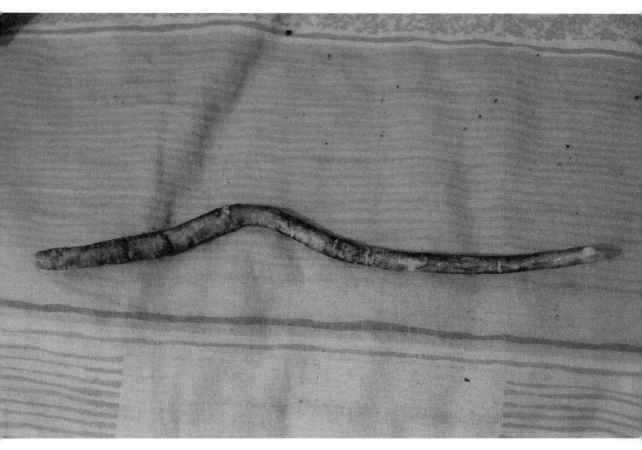

Elagaq or *qerqaq* (an "Eskimo potato"), the root of alpine sweet vetch. KEVIN JERNIGAN

branches; and some are thick. Those are different sizes. They are long and short. . . . They add them to various things, they make soup with them."

Nasqupaguat
Nutty saw-wort

Nasqupaguat (nutty saw-wort, lit., "pretend *nasquq* [head]," *Saussurea nuda*) are found in wet areas close to the coast. Theresa Abraham (February 2019:22) explained: "They grow down in the marshland on the coast along rivers. They have small leaves like little *quagcit* [sour docks]." People cook the young leaves with fish in spring, before they harden. The greens can also be eaten raw with seal oil. Neva Rivers (May 2004:86) said that she added *nasqupaguat* to fish starting in May when they had just started growing.

Nelson Island elder Albertina Dull (November 2007:315) recalled her mother gathering them along the shores of sloughs. Martina Aparezuk (December 2012:72) of Kotlik noted: "I also sometimes gather those that we call *nasqupaguat* once in a while and prepare them in various ways. They are delicious." Cecelia Andrews (April 2019:240, 104) added:

> I eagerly pick *nasqupaguat*. Their large tops, their heads, are their males. And they also have females. They mostly pick their males. . . .
>
> They are very delicious as an ingredient added to *akutaq*. . . . After gathering some, we'd cook them, and when they cooled, we would cut them into pieces and make *akutaq*.

Muugarliarniaret
Roots of water weeds

Muugarliarniaret, the roots of *nuyaruat* (water weeds, lit. "pretend hair," *Sparganium* spp.), are also edible. Francis Charlie (November 2014:116) explained:

> In spring, after breakup, the roots of [*nuyaruat*] that the ice broke off float to the surface. [Ice] that floats away probably pulls them out. They call [their roots] *muugarliarniaret*.
>
> They are very white. . . . The tips of the male [plants] are sharp. But the tops of those that are apparently their females aren't sharp, and they taste like sugar.

Sparganium species (known locally as *nuyaruat* or "water weeds") cover the surface of a coastal pond. The plant grows in beautiful rings, evident from the air. Notice the swan-shaped silhouette formed by the water weeds, and the tiny white dots indicating that a swan family is swimming (and eating) among them. J. J. FROST

Francis noted that moose, as well as swans, northern pintails, and greater scaups, eat water weeds in the fall to fatten up, and he eats their roots:

> They are delicious. I sometimes gather those at our camp [near Black River] in spring. And my younger children and grandchildren have gotten used to eating *muugarliarniaret* as well. . . .
>
> They grow in rivers, and we call their tops *nuyaruat*. Some of them are long grasses. Their males are like green grasses, and their females have leaves. . . . I just remove those sharp parts and eat them. They are their males, as they say. . . .
>
> Right after breakup, [their roots] beach on the shores up around my camp. And there at Anagciq, I would go down to the shore and eat for a long time. [*chuckling*]

Qatlinat wall' qacelpiit
Stinging nettles

Qatlinat or *qacelpiit* (stinging nettles, *Urtica gracilis*), both meaning "those that sting," are plentiful in southwest Alaska in wooded areas, and they are harvested in early summer. John Andrew (April 2019:267) recalled:

> They say they pick these when they're young and cook them. Before they harden, they pick them and add a lot of them when they make fish soup, when they make salmon soup.
>
> They have needles, that's why they call them *qatlinat* [lit., "those that sting"]. When it has hardened and you touch it, [the skin] turns red and swells.
>
> They call them *mingqutnguayaalget* ["ones with small *mingqutet* (needles)"] and *qatlinat*, too.

Peter Gilila (February 2019:108) of Akiak and Nick Andrew recalled how nettles were used to make twine and fish nets in the past. Nick said: "They strip them and twist them; they said they're strong and durable. They used to make those into gill nets."

Iitaat
Tall cottongrass

As summer deepens, people continue to harvest plants. *Iitaat* (tall cotton-grass, *Eriophorum angustifolium*) are gathered in July, when their lower stems have begun to fatten. Marie Alexie (June 2016:231) of Atmautluak recalled: "They gather them when they start to get fat in summer during the berry-picking season. They pull them, but don't [pull out] their roots. We also dip them in seal oil and eat them with other things." George Pleasant (January 2011:270) said: "When it was time to pick *iitaat*, [my grandmother] would pick them. *Iitaat* are good to eat by just removing their outer peels and dipping their ends [near the root] into seal oil."

While the lower stems are eaten, the grass itself is gathered and dried for use in braiding fish and twining large grass storage containers. Theresa Abraham (November 2007:312) noted: "They gather those *iitaat* down on the coast. Then they remove the edible part [lower stem] and soak it in water, and

> *Qatlinat* (stinging nettles), also known as *qacelpiit*. KEVIN JERNIGAN

Iitaat (tall cottongrass), July 19, 2014. KEVIN JERNIGAN

then they eat them by dipping them in seal oil. Then they used [the grass section] for braiding [fish], or they also used them for weaving and twining in the past." Examining a photograph of a woman carrying a bundle of *iitaat* in the 1930s, Frank Andrew (February 2003:841) commented: "They never went without these *iitaat*. And their peels are medicine for the inside of the body. We eat them as well after dipping them in seal oil. And they use their tops, weaving them into *kuusqulluut* [grass storage bags] for fish. They use *iitaat* and *kelugkat* [water sedge]. They never went without these."

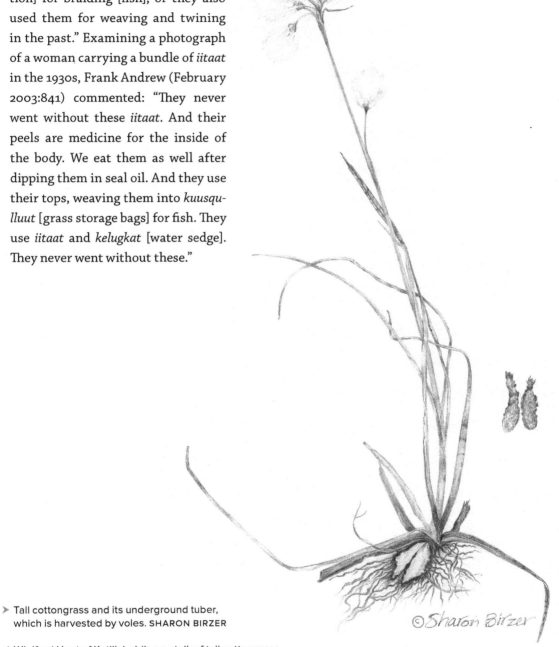

> Tall cottongrass and its underground tuber, which is harvested by voles. SHARON BIRZER

◄ Winifred Hunt of Kotlik holding a stalk of tall cottongrass, the lower part of which can be eaten raw with seal oil. JACQUELINE CLEVELAND

Pingayunelget
Marsh fivefinger plants

Pingayunelget (marsh fivefinger plants, from *pingayunelgen,* "eight," *Potentilla palustris*), also known as *mecungyuilnguq* (from *mecuq,* "liquid part of something, juice"), grow among bushes and trees in southwest Alaska. Nick Andrew (February 2019:71) noted that both moose and caribou find them tasty, but few recognized them as edible. Cecelia Andrews (April 2019:109) briefly described their use:

> These *pingayunelget* that I tell people about, our father's grandmother, Marpak' [picked them]. . . . They would pick their tops and make tea and drink tea. *Pingayunelget* [marsh fivefinger plants] were their tea.
>
> They used them, even though they weren't dry. They are along the shores of lakes on high tundra. When you pick berries you will see them.

Pingayunelget (marsh fivefinger plants). KEVIN JERNIGAN

Kulukuunaruat (bluebells). KEVIN JERNIGAN

Megtat neqait wall' evegtat neqait
Bumblebee food

A number of edible blossoms go by the lovely descriptive name *megtat neqait*, or *evegtat neqait* (lit., "bumblebee food"). John Andrew (April 2019:316) noted yet another name: "The flowers in the wilderness around our area and on the marshlands, they are food for *panayulit* [bumblebees]." Cecelia agreed: "They eat any kind of flowers, anything sweet."

Jernigan (et al. 2015:143) notes an old spelling for bluebells (*Mertensia paniculata*) as *punaiyulinu'kait* (*panayulit neqait*, "bumblebee food"). Also known as *kulukuunaruat* (lit., "pretend bells"), children ate bluebells as they played in summer. John Andrew (April 2019:183) remarked: "Those bluebells are plentiful far upriver here in fish camps. When they grew, children would eat their flowers, saying they tasted sugary." Ruth Jimmie agreed: "We used to eat those, too."

Working in Tununak in the late 1970s, Thomas and Lynn Price Ager (1980:39) wrote that children sometimes ate the nectar-rich flowers of several species of *Pedicularis* (woolly lousewort) that grow on Nelson Island: "All these local species, as well as several other insect-pollinated tundra flowers, are referred to as "bumblebee flowers."

Quagcit quunarliaraat-llu
Sour dock and mountain sorrel

Quagcit (sour dock, *Rumex arcticus*), also known as *quunarlit* (from *quunarqe-*, "to taste sour") in Kotlik and the upper Kuskokwim, is well known and widely used throughout southwest Alaska. Jernigan (et al. 2015:89) notes that in Hooper Bay and Chevak, the leaf is sometimes referred to as *aatunaq* and is considered female, while the flower stalk is referred to as *naunrayagaq* (lit., "small *naunraq* [plant]") and is considered male.

John Andrew (April 2019:82) described harvesting them in mid-summer after their stalks had turned red:

> Those older than us would ask us when we arrived, "Have the tops of *quagcit* not turned red?" At first, I didn't think anything of it. Then one day I asked someone, "Why do you always ask whether or not they have turned red?" He said when they turn red, when they get sour, they become delicious. They'd tell us not to gather them before they turned red. Only when their tops started to wilt and they turned red, they said it was time to pick them.

Sherrie Heakin harvesting sour dock leaves, July 13, 2019. JACQUELINE CLEVELAND

◄ Sherrie and Kristen Heakin with
  bags full of sour dock leaves.
  JACQUELINE CLEVELAND

◄ Sour dock plants.
  JACQUELINE CLEVELAND

▽ Sour dock plants.
  SHARON BIRZER

©Sharon Birzer

In the past, people harvested and stored large amounts of *quagcit*. John (April 2019:83) continued:

> My late wife [from Hooper Bay] told me that when they were small, they would gather a large amount of these. And then after cooking them, they stored them inside large wooden barrels.
>
> She said after putting other plants over them to cover them, they would keep them in their partially underground storage caches. She said when they wanted to make *akutaq*, their women would go down and after taking enough to make *akutaq*, they'd cover them again. She said ones that are stored in wooden barrels last a long time.

Cecelia Andrews (April 2019:83) agreed: "Some people kept them in their partially underground storage caches; they had sod [caches] lined with wood inside. You know they had fridges, and stored food wouldn't spoil."

Quagciq leaves are commonly eaten raw with seal oil and a little sugar, or cooked and mixed with berries to make *akutaq*. Martha Mark (January 2011:273) recalled:

> When we picked berries, we had small wooden barrels for containers. We'd fill them with salmonberries, and then we'd cook some sour dock, and when it was thoroughly cooled, we added it to the salmonberries, placing it along the bottom and not letting it be visible, placing it along the middle of the salmonberries.
>
> Then in winter back when there weren't freezers around, it froze. And then when I was about to make *akutaq*, I would get some berries to add from there. Adding sour dock to the salmonberries tasted good. I used an *uluaq* [semi-lunar knife] [to get some berries out], and then when it melted, we'd mix it into the *akutaq*. It was sour dock and salmonberries mixed together.
>
> Now that we've gotten freezers today, when I gather sour dock, I store them separately after cooking them. Then I add salmonberries to them and make *akutaq*.

Martina Aparezuk (December 2012:71) described cooking sour dock leaves, then cutting them into small pieces and mixing them with blackberries or blueberries. This mixture, known locally as *atsaarrluut*, can then be used to make a tasty *akutaq*. Other greens, such as *nasqupaguat* (nutty saw-wort), can also be added to the mixture. Sour dock tops, sometimes referred to as *nakaat*, can also be boiled, then strained and eaten with seal oil and sugar,

Quunarliaraat (mountain sorrel). KEVIN JERNIGAN

providing a stronger-tasting but delicious treat.

Theresa Abraham (July 2007:239) remembered her mother using sour dock leaves to line large, open-weave baskets to store berries:

> Since they had no freezers in the past to store them, since these black-berries tend to turn a light color when they are kept on the ground [they stored them underwater]. Inside a *naparcilluk* [grass storage basket], starting along the bottom, she would place *quagciq* inside there to line it, and she'd fill it with those blackberries. . . . And she'd sink it down in the water along the edge of a lake and keep it underwater.

Jernigan (et al. 2015:90) notes that some consider cooked sour dock leaves and stems, mixed with either seal oil or cranberries, as a treatment for intes-tinal parasites. Martina John (November 2007:306) stated that on Nelson

Island *quagcit* were considered primarily as food: "Before their males wilt, before they become tough, they are good when cooked. And we don't consider them to be medicinal at all, but they are our food, we are very eager to eat them." Helen Walter (November 2007:315) has cultivated *quagcit* outside her home in Tununak: "I brought them down with their roots and planted them. The ones that I planted last year grow during summer."

Nelson Island women also mentioned *quunarliaraat* (mountain sorrel, lit., "small *quunarlit* [sour dock]," *Oxyria digyna*), which they called "the other type of sour dock" or "the cousins of sour dock." Their leaves are used in the same way as *quagcit*. Martina John (November 2007:307) noted: "They are smaller and extremely sour. They call those *quunarliaraat*."

Angukat wall' nakaaret
Wild rhubarb

Angukat (wild rhubarb, from *angun*, "male," *Polygonum alaskanum*), also known as *nakaaret* and *arnaurluut* (lit., "poor, dear *arnat* [women]"), are widespread throughout southwest Alaska, though not as plentiful around Nelson Island. Leaves are eaten raw, dipped in seal oil, and both leaves and stems are sometimes cooked. Paul John (December 1985:135) noted: "They eat the leaves of *angukat* and their stems. And when they cook their leaves and store them, they won't go bad. And they also mix them into *akutaq*." Wassillie B. Evan added: "We also eat them as they are, adding a little processed salmon eggs, mashing them, and adding a bit of seal oil. Those were just like 'pie' for our ancestors." According to Peter Jacobs (October 2003:123) of Bethel, wild rhubarb is the mate of *quagcit* (sour dock): "They say that *quagcit* are the females to *angukat*." John Andrew (April 2019:77, 79) noted that *angukat* were taken care of in the same way as sour dock:

> Before their tops grow, when their leaves get big, they gather them in large numbers. And then they cook them like *aatunat* [sour dock], and make them into *akutaq*.
>
> And before their stems get too hard, they peel them and eat them. And some people dip them in sugar and eat them.
>
> Even though their tops have grown, some people pick them. They remove their leaves and cook them. My late wife, after she had gotten a lot, she'd cut them up and then cook them, but she wouldn't cook them for long. . . .

> ➤ *Angukat*, also known as *nakaaret* (wild rhubarb)
> KEVIN JERNIGAN

Wild rhubarb. RICHARD W. TYLER

You can freeze it after you cook it. If you freeze it just the way it is, it
will just crumble and wilt away. . . .

If you cook it inside the house, your house will smell like sourdough.
[*chuckles*] Instead you gotta cook it outside or in your smokehouse.

John said that when smoking fish, wild rhubarb leaves can be added to the
wood to increase the smoke and hasten the process. He also noted that wild
rhubarb grows alongside rivers, and can regrow after being uprooted: "In
some fish camps upriver around our village, when the people who had them
threw them there, they grew again."

Peter Jacobs and Frank Andrew (October 2003:123) recalled the saying
that in cold, windless weather, wild rhubarb would appear as tall as a person. Peter posed the question: "What about when the wind is always calm
in winter? I think there is a saying about that. You know *nakaaret*, the sour
dock's mate, would get as tall as a person because there would be no wind
[and much frost]. . . . You know when the wind has been calm for so long. Is
that saying true?" Frank Andrew replied:

They say when the world was in its original state and was cared for,
and not contaminated, that was how the world was. . . . They say that
in January, when the weather was in its original state, the wind would
weaken. And the *nakaaret* would get as tall as people because there was
so much frost. The shorefast ice on the ocean would be far out near the
edge of deep water. That's why those people apparently were successful
catching animals when they hunted.

Tayarut
Mare's tail plants

Tayarut (mare's tail plants, *Hippuris tetraphylla*) are another versatile and
well-known plant. *Tayarut* grow in ponds all over southwest Alaska. People
usually gather *tayarut* in the fall after lakes freeze. George Pleasant (January
2011:272) noted: "After it had frozen, when winter came, the last things to
gather were *tayarut*. [My grandmother] also gathered *tayarut* from lakes that
had just frozen. They gather them after they wilt." Paul John (December
1985:51) described harvesting *tayarut* in the fall:

Tayarut grow in lakes. . . . When the ice is thick, if we can find them we

Tayarut (mare's tail plants).
KEVIN JERNIGAN

were told to put away as much as we could.

You'd gather them and put them away and leave them there. They would keep though they didn't use them, and when their food supply ran out, they'd cook them, adding a little bit of food. They are nourishment when you add a little food taste to it.

Wassillie B. Evan (December 1985:51) said that *tayarut* remain erect in lakes after freeze-up: "They would remain standing and not fall to the side." People tried to harvest *tayarut* before the first snow. Martina John (November 2007:448) observed: "Sometimes when [the snow] covered them, it was difficult to obtain them." Theresa Abraham (November 2007:448) explained:

> When it has snowed on the water before freeze-up, the conditions aren't good [to gather *tayarut*]. Then when it freezes, the lake appears as though it has foam on top of the ice and it isn't smooth, and that covers the *tayarut*. That's what happens down on the coast. . . .
>
> But if it had frozen without having snowed, those *tayarut* are in very good condition.

Martha Mark (January 2011:274) said that *tayarut* could also be gathered in spring along with *kapuukaraat*: "I [also] obtain *tayarut* when they grow [in spring]. In lakes, they are green and just starting to get long. . . . I also obtain them like [George] mentioned in fall when they've wilted. And in spring again, when [lakes] melted, we obtain *tayarut* once again. When we obtained *kapuukaraat*, we obtained *tayarut* in the past, ones that had grown the previous year." John Andrew (April 2019:25) described harvesting *neqnirliaraat* (the roots of *tayarut*), brought to the surface when the ice along the shores of lakes broke up in spring: "When it started to get warm before the fish run, we would get *tayarut*. And when there was water from melted snow, when

Kristen Heakin holding a stalk of mare's tail, July 31, 2019. JACQUELINE CLEVELAND

Tayarum iluraa (mare's tail's cousin, *Hippuris vulgaris*). KEVIN JERNIGAN

neqnirliaraat came to the surface along the shores and swans started to eat them, they let us gather some that we would eat right away. They were delicious, they tasted sugary."

Elders noted different harvesting methods. Simeon Agnus (July 2007:510) of Nightmute explained: "During winter right after freeze-up, we would collect them by scraping them [off of the glare ice] with a shovel. One has to make an effort at having a supply." Some use rakes to gather *tayarut*. Mike Andrews (December 2011:485) of Emmonak recalled: "When we arrived [at the lake], after gathering them with a rake from down there, after bringing them up, we filled our bags. Using a rake, we waded farther down and brought up more. I obtained some, filling that black plastic bag." Mike's wife, Maryann Andrews (April 2012:75), had used a sled: "I just take the sled and [push the runners sideways against the plants]. Many would fall, and they would pick them. They are very tasty." Not only adults gathered *tayarut*. David Martin (May 2004:73) recalled: "They asked boys to gather *tayarut* when it was time, since we were told that the plants that grow on land are food. And I used to gather them."

Tayarut were relatively easy to harvest and were gathered in large quantities along the coast. Neva Rivers (May 2004:73) explained:

> My mother and I used to gather them in large numbers from down
> in the marshy area. And then we would go back and forth to get them.
> And since another person wouldn't fetch them, I would use a kayak sled,
> filling a woven grass container to get the ones we had gathered.
>
> During winter, when they asked me to get some from there, I went and
> got them using a kayak sled, not with a snowmachine. I enjoyed fetching
> them.

Pauline Jimmie (March 2007:71) of Chefornak spoke of the importance of *tayarut* in times of food shortage:

> My father told me that I should always gather a lot of *tayarut* during
> fall. . . . He said that *tayarut* mixed with seal oil is a food that can be
> eaten during starvation times. He said long ago when they experienced
> famine they ate *tayarut*. My grandmother fed those people *tayarut*, but
> mainly with seal oil in it.
>
> He told me that I should gather *tayarut*, even if it was a large amount,
> that it would be okay if I discarded the leftovers during spring when
> birds were around. That's why I always try to gather *tayarut* during fall
> right after freeze-up. That person and I, with each other's help, always

try to fill two plastic bags. I discard them during summer when I know that we won't go hungry.

Paul John (December 1985:132) described the importance of teaching young people today about *tayarut* in case of emergency: "When we were talking about the land's plants and berries, *tayarut* and lichens, because they are available now, if we let someone cook them, not just telling our young people, or if we let them eat them, later on if they are lost and run out of food, that possibility would be open to them."

Tayarut were both plentiful and sustaining in times of need, and people prepared them in a variety of ways. Paul John (September 2003:97) declared: "These *tayarut* are used in everything that is cooked." Earlier, Paul (December 1985:132) had summarized this variety: "*Tayarut* have many uses. You can use them with sea mammal blood, adding them to the broth. . . . Anything that has blood, moose or caribou or rabbits, they mix *tayarut* into those. Also whitefish or burbot eggs, they add [*tayarut*] to their broth." Wassillie B. Evan (December 1985:132) noted that his friend's sister mixed seal blood with *tayarut* and that they were delicious: "*Tayarut* were used just like rice by our ancestors, adding it to soup." Paul John added: "Still at this time we haven't stopped using them as rice. We would cook them with burbot eggs and eat the soup with great enjoyment."

Elders from throughout southwest Alaska described adding *tayarut* to broth soup. George Pleasant (January 2011:272) recalled: "Some people, when they return home, they make broth soup, *qayussaagluteng* as they say, filling the pot with salmon roe and putting it on the stove, and they add a certain amount of aged salmon roe to it and make soup, placing *tayarut* in it." Martha Mark (January 2011:272) also used *tayarut* for broth soup: "After cooking bird without adding rice to the broth, I scoop [the bird] out. Then afterward, I put a little bit of aged salmon roe in it, and then I put some *tayarut* in. It's very delicious as broth soup."

Anna Agnus (July 2007:510) of Nightmute described *tayarut* as *cuassaat* (wild greens that can be cooked):

They also gathered [*tayarut*] when they were newly grown. They prepared them as *cuassaat*, cooked them, and ate them with seal oil.

Even in summer, they would add them to seal meat they cooked, preparing them into soup.

Anna's husband, Simeon Agnus, noted:

They add fish innards to them and make broth soup out of these. . . .

During winter, when we make soup out of them, they are very tasty, but you have to add some fish taste to them. If they are placed inside fish broth where fish was boiled, they become edible.

Anna continued: "They also kept fish eggs and seal blood because of these plants. There are many foods to eat on the land, and many women know what they are since they are cooks. You know how white people add various things when they make soup. That's how these [plants] on the land are."

Mark John (April 2012:75) of Toksook Bay made his audience chuckle when he recalled yet another variety of broth soup: "My wife calls them sperm soup when we add burbot sperm and eggs to them. My younger sister started [calling them that]. Then my wife, when we cook those, calls them sperm soup." Maryann Andrews concluded: "They are good in any kind of soup, even snowshoe hare [soup]."

Theresa Abraham (April 2019:56) said that when adding *tayarut* to broth soup, the soup is done when the *tayarut* sink. Martina Wasili (March 2008:223) from Chefornak noted that in her hometown if one is cooking meat and adds *tayarut* to it, everything will get done cooking at the same time. Others agreed that *tayarut* from the coast cook faster than those harvested farther inland. John Phillip (February 2006:134) noted: "The *tayarut* in lakes vary. Some don't cook all the way through and stay firm. And some cook very easily."

Along with soup, *tayarut* can also be eaten fresh or made into *akutaq*. Theresa Abraham (April 2019:54) explained:

Our parents used *tayarut* extensively, adding them to their cooking and eating them. And then, in summer when they started growing, we'd really eat them. When they didn't add it to their cooking, they would eat them with a little seal oil.

And then more recently, some people have started to cut their tops and cook them and make them into *akutaq*.

Several mentioned a closely related plant, *tayarum iluraa* (lit., "mare's tail's cousin," *Hippuris vulgaris*). Theresa Abraham (April 2019:162) described them as looking like *tayarut* but with smaller leaves. John Andrew remarked: "They also call these thicker ones *tayarut* around here. . . .Those skinny ones grow mostly in slightly deeper water, and those thicker ones in shallow water." Cecelia Andrews added: "They called those larger ones *tayarulungniit*."

Neqnirliaraat (oysterleafs). KEVIN JERNIGAN

Neqnirliaraat
Oysterleafs

. . . .

Neqnirliaraat (oysterleafs or beach bluebells, *Mertensia maritima*) were rarely mentioned by the elders we worked with, although their name—which translates "best-tasting things"—implies that they were a delicacy. Ager and Ager (1980:38) wrote that one Nelson Island woman reported collecting them before they flowered, cooking the stems briefly, and eating them with seal oil. Jernigan (et al. 2015:76) noted that Mary Pete of Stebbins said that the root was used as a lure for fishing, as its distinctive odor attracts fish.

Neqnirliaraat (oysterleafs). KEVIN JERNIGAN

As noted above, John Andrew (April 2019:25) used the name *neqnirliaraat* for the roots of *tayarut* (mare's tail plants). Nick Andrew (February 2019:9) also talked about *neqnirliaraat* as a lake plant: "Then there are *neqnirliaraat*: we have a number of lakes [that have them]. They say those *neqnirliaraat* come to the surface in the fall. The edge [of the lake] has those beached. Birds that eat from there get fat right away. They are very delicious." That the same Yup'ik name is used for two different plants is not surprising, as the name is descriptive of taste, not appearance.

Wild rose bush. KEVIN JERNIGAN

Tuutaruat
Rose hips

Tuutaruat (rose hips, lit., "imitation labrets," *Rosa acicularis*) were gathered in the fall and either used for tea or cooked and added to *akutaq*. John Andrew (April 2019:99) explained:

> They are abundant farther upriver on the Kuskokwim. They grow in places with a lot of trees. . . .
>
> [They pick them] when their flowers have wilted, when their *tuutaruat* start to rot, when they get soft after turning red. When picking them when they're green, people say they cause stomachaches.
>
> They make them into *akutaq*. And they boil them to make tea.

A *tuutaruaq* (rose hip). KEVIN JERNIGAN

My wife would also add them to salmonberries, other berries, and even with crowberries and make them into *akutaq*. She made *akutaq* out of all sorts [of berries and plants].

John (April 2019:101) then told a story about his late wife picking *tuutaruat*:

When she learned about these, she liked picking them. She would bring a dog with her, and she'd tell me that she was about to go and pick *nangengqauryarat* [ones that require standing (when picking)]. I would smile at her.

Then I asked her one day, "Why do you call them that?" She said in her hometown [Hooper Bay] there are no trees that require standing [when picking].

When she'd call them *nangengqauryarat*, I'd smile at her because we don't call them *nangengqauryarat*. She would tell my family that she picked *nangengqauryarat*. [*chuckles*]

Urut
Sphagnum mosses

Urut (Sphagnum mosses, *Sphagnum* spp.) are widespread in tundra areas of southwest Alaska and had a variety of uses. Mosses could be used as wiping material, pot scrubbers, bandages for cuts, and lamp wicks. Wassillie B. Evan (September 2003:95) said that if people were short on food, they made soup out of moss to feed their dogs. Margaret Andrews (December 2002:167) of Kotlik noted that women dried large quantities of *urut* in the fall and used it for lining diapers as well as for sanitary pads:

> When they are pulled out they are kind of long. They dried those in the fall for winter. When I was becoming aware, those who lacked used those, and they would use them in their diapers. Those who had their periods also used that. That was before Kotex became available. Moss is still in use. When hair seals are skinned, mosses are used to wipe up the blood.
>
> We used to put them in sacks. That's what my grandma used to do. They didn't lack all winter long. You know, I still use mosses to wipe things. They're like sponges. [*laughs*]

Marie Alexie (June 2016:215) described using mosses to cover fish when making *kumlanret* (raw frozen fish). Other ingenious uses were as *kumarutet* (moss wicks) for seal oil lamps, as fire starters, and as nightlights or candles when going between houses. Theresa Moses (September 2003:95) of Toksook Bay shared: "They took out a large piece and dipped it in seal oil. Or if they went to the food cache [at night], they used it as a light." Albertina Dull (November 2007:4-5) remembered:

> They'd place a piece of wick in a dish or small clay [lamp]. Back when cloth was scarce, we used those ones from the tundra.

Urut (Sphagnum mosses). KEVIN JERNIGAN

Urut braided and dried for use as *kumarutet* (moss wicks) for seal oil lamps. One-hundred-year-old Albertina Dull created this braid to show us how it was done when she was young. ANN FIENUP-RIORDAN

They called them *kumarutet* [lit., "things used to light (fires)"]. They gathered them and dried them. Then they removed one piece that was a good size and used it as a wick to light a flame. There was nothing else, and matches were extremely scarce back when I became aware of my surroundings.

And they kept their lamps on for a long time. They never kept stoves on all the time. When they made boiling water with a kettle, if they had a small amount of tea, they tried not to let the lamp's flame go out.

And if there was someone cooking, as they only cooked outside in summer when they stayed in tents, if a person was cooking outdoors, by placing a hearth around it, placing the kettle or pot on a wooden hanging device, after taking that moss wick and placing it in some seal oil, we'd get something to block the wind, and we'd go to that flame that was burning and light [the moss wick]. We were getting flame, *kenerciluta* as they say. That's what they used to do. That's how those poor people were.

Elsie Tommy (August 2012:29) of Newtok spoke of seal oil lamps with moss wicks made of dried *urutvaguat* (lit., "big, pretend *urut*"), which she noted are different from *urut* and which may be juniper polytrichum moss (*Polytrichum juneperium*) which grows on peat and is somewhat flammable:

They are dark and look like small mosses and are long and thin, and the small things are tufted. They would gather large amounts of those in summer and dry them and use them as wicks on their lamps all winter back when Western goods were scarce.

They aren't *urut*, they are *urutvaguat*. They are plants that easily ignite. When pulling off [their tops], inside them there are very small tufts. There are many of those growing down on the coast. We didn't gather them from around places where there were bushes growing but on the tundra. They look a little greenish. They grow close to *uruq* [moss]; they are called *urutvaguat*.

Reindeer only eat the other kind, the *qilungayagaat* [probably reindeer lichens, from *qilu*, "intestines, entrails"] that are white that lie on the ground.

Perhaps the most memorable use of *urut* was as *puyat* (mosses soaked and aged in seal oil), which could be used both to caulk a kayak or mixed with berries to create a special mixture, which Annie Jackson (July 2000:29) of Akiak noted cleansed one's insides. *Puyat* could also be eaten in an emergency as "famine food." John Andrew (April 2019:294) explained: "Back when they used to go through starvation, they said they could soak them in seal oil and eat them after aging them. They said a person won't starve when eating these. . . . For us much later on, when we want things, we turn to the stores. . . . Our ancestors used everything in the past."

Lizzie Chimiugak (January 2007:589) described the process of making *puyat*:

Long ago our dear ancestors regarded [*puyat*] as essential commodities. . . . They were never without *puyat*. They put *puyat* in seal stomachs and let them soak in seal oil. . . . I first saw *puyat* in seal-stomach containers when plastics were not readily available.

Nuuniq [Louise Ayaginar] taught me how, taking a *puyaq* like this. How did they eat these when they didn't make them into *akutaq*? When she did [eat some], she plucked some down feathers from my parka, and she told me to mix them [with *puyat*] and chew on it, and she also chewed. She said that they ate *puyat* like that long ago when they'd go through starvation.

Sophie Agimuk (January 2007:590) of Toksook Bay continued: "They used to make *puyat* using those *urut* [mosses] they had collected. . . . After they picked them, they dried them. And then when they dried, they soaked them with new seal oil . . . and put them away. And then those pieces of moss get so saturated with jelled oil that they become thick and sticky. That was the way they made [*puyat*]."

Puyat could be left in the sun, and the resulting sticky mixture was applied to the seams of kayak and boat skins to make them watertight. *Puyat* was also used to make a special *akutaq*. Sophie Agimuk continued:

> When they mixed them, they kept mashing it by adding water to it. Soon that mixture would multiply. When it became enough, they added blackberries. . . .
> But as soon as one finishes making *akutaq* out of it, it must be eaten right away. If it is left alone, it becomes watery and the berries inside it come to float on top.

John Alirkar of Toksook Bay noted: "*Tengluni-gguq tua-i* [So it 'takes wing' as they say]." Lizzie Chimiugak continued: "They said not to leave any of it uneaten; they said that it would fly away. And believing them, I thought that it would be floating up there."

Phillip Moses, also from Toksook, shared the story of his older sister's experience when making *puyaq akutaq*:

> I watched my older sister, Arnaqulluk, down there at Qalulleq when she made that kind of *akutaq*. This was long before she got a husband. . . .
> So when she added berries to it, she found that they were not enough. So covering her *akutaq*, she hurried up above Qalulleq to get more berries since there were usually blackberries there . . . to put in [her *akutaq*].
> So when she came in she went to her *akutaq*. It had become nothing but water! [laughter]
> It had "flown" as they say! Probably because she had seen that happen for the first time, suspecting her sisters and also the father of Cacungaq [she exclaimed], "Now who in tarnation poured water into my *akutaq*!"
> So my mother told her that since she had left it to get berries to mix into it, it had flown.
> Sometimes I think of that, when she did that in my presence.

Cecelia Andrews (April 2019:288) had been warned not to make noise while her mother made *puyaq akutaq* lest it fly away: "They used to scold me,

'Gee, you there, don't make noise at all.' She would say her *akutaq* would fly away if we made noise. [*laughter*] When we were small, that's the admonishment we were given. Gee, looking [at the *akutaq*], it was very smooth. They said indeed, when they fly, [the *akutaq*] would get lower and even melt."

Finally, *urut* could sustain a person in other ways. John Andrew (April 2019:291) described his experience when caught far from home:

> I stopped, since it was getting dark, and I happened to see a large spruce and small trees around it. Since the area was dry, I built [a shelter] by binding their tops. And after gathering some mosses from there, I put a roof over it, I insulated it. And after getting some [spruce] branches, I also put some inside and put moss on top of it, making a mattress. . . .
>
> I kept the fire on all night. And after getting some mosses, I used them for a blanket. . . . My body was warm. Mosses saved my life.

Tuntut neqait, ciruneruat-llu
Reindeer lichens and kidney lichens

Several varieties of lichens are also abundant in the tundra regions of southwest Alaska and are referred to by a variety of names. Reindeer lichens (*Cladonia rangiferina*) are known as *ungagat* on Nelson Island (lit., "pretend *ungiit* [whiskers]"), *taqukanguat* along the Kuskokwim (lit., "pretend *taqukaq* [brown bear or seal]"), as well as *tuntut neqait* (lit., "reindeer food"), a name that was also sometimes applied to Sphagnum moss. Kidney lichens (*Nephroma* spp.) were known as *ciruneruat* (lit., "pretend antlers") along the Yukon River.

In the past, people ate lichens. Grace Parks (June 2016:231) said she added lichens to soups and ate them fresh as well: "I really like those they call *tuntut neqait* [reindeer lichens]. I eat some on the land." Paul John (September 2003:95) recalled: "Before my late grandmother died, I used to eat those soups made of *ungagat*. After my grandmother passed away, I didn't eat them again." Like mosses, lichens could also be soaked in seal oil and eaten in an emergency as "famine food." Both Albertina Dull and Lizzie Chimiugak (May 2019:74) mentioned eating *ungagat* during starvation times. Lizzie added: "We can eat *ungagat* like *tayarut* if we don't have any food."

Nick Andrew (February 2019:7) described eating lichens to stave off hunger when traveling in the wilderness: "They also instructed us that if we don't have anything in our stomachs out on the land and we are unable to get anything, we should go on the tundra and those [lichens] that the

Tuntut neqait (reindeer lichens). KEVIN JERNIGAN

caribou used to eat that are white, they are reindeer food. They said if people eat those, even in small amounts, they make one feel better." Theresa Abraham (April 2019:280) agreed: "I heard that if someone was walking in the wilderness and didn't arrive, he could eat some *tuntut neqait*, even if it's not a lot, to put some food in his stomach." Ruth Jimmie (April 2019:282) recollected: "Mr. Beaver who arrived in your village [Chefornak] had survived, remembering what his grandfather said. He knew what to eat."

Peter Gilila (February 2019:101) added that lichens should not be eaten dry but only after soaking them in water. Peter (February 2019:11) also said that lichens could be added to dog food when supplies ran low in spring: "When we started to run out of things, including dry fish, they'd even have me gather those, and they would include them when they made cooked dog food. We'd also eat them."

Palurutat
Edible mushrooms

Susie Walter (April 2008:215) remembers gathering *palurutat* (edible mushrooms, possibly from *palurte-*, "turned over or belly-down") when berry picking: "When they are new, when we go and pick blackberries, we eat them along with [berries]. We add the berries to them. And here when they become [mature], we run away from them, thinking they become poison." Though some people consider mushrooms inedible, others remember them as tasty. Ruth Jimmie (April 2019:296) also recalled picking mushrooms, filling them with crowberries, and eating them as snacks out on the tundra. Cecelia Andrews (April 2019:296) remembered eating brown mushrooms: "When you put them in with salmonberries for a while, they really soak in the juice. They are good eating." John added: "After putting some blueberries on it, we would eat it like a sandwich."

John Andrew (April 2019:295) said that people picked the larger mushrooms—probably boletes (*Boletus edulis*)—and used them to cover the contents of berry buckets:

Palurutat (edible mushrooms, possibly *Boletus*) growing near Scammon Bay. JACQUELINE CLEVELAND

When we were picking berries, the women who weren't young would put the large [mushrooms] on top of the berries they picked. . . .

And they could cook them, they could add them to soup, they could fry them.

But they'd also tell us, "Don't try the ones you don't know, that you don't recognize."

This is good advice. Young people are admonished to gather and eat only plants that knowledgeable men and women indicate are safe to consume.

Qecigpiit
Unidentified aquatic plants

Albertina Dull (November 2007:309) described plants that she called *qecig-piit* (lit., "ones with big *qecik* [skin]") and which we cannot identify:

You don't know *qecigpiit* because you haven't seen them before. Those plants that grow along saturated ground. Those that are nothing. They are probably not medicinal plants. . . .

Those that are newly grown make crunching noises when you eat them, and they are good eating.

Helen Walter added that she gathers *qecigpiit* on mountainsides: "They are in water, and they are thick and green. I happened upon some and ate a lot. . . . When I gather *quagcit* [sour dock] from the top of our mountain, I always search in ground saturated with water. When climbing up, there is a lot of *quagcit* near our village [of Tununak]." Lizzie Chimiugak (May 2019:175) also mentioned *qecigpiit*:

Qecigpiit were also delicious. They are in springs in the mountains and are green. We would take them from the water.

And when people and I ate those *qecigpiit*, they were almost like Jello. . . . They are in springs up on the mountain; they float in small springs.

Ariqat
Type of sand dune plant

Finally, Nelson Island elders gave vivid descriptions of *ariqat*, a plant found in their homeland but which we are unable to identify. Albertina Dull (May 2019:171, 212) recalled: "The *ariqat*, people would take the tops and pull them from the land. People would chew and swallow them. . . . Only people with good teeth chew and eat them." Her younger sister, Lizzie Chimiugak (May 2019:174), continued:

> There are *ariqat* right above my fish tending place. . . .
> They have white flowers on the top, and their roots are large like fingers. When you gather them, they are delicious, and they seem like they would be good dipped in seal oil.
> [Their roots] are not the same size when dug up. They are good to chew after washing them.

Later, Lizzie (May 2019:178) continued:

> My mother used to tell stories about food shortages a long time ago. . . . When the ocean down there was not open, there would be no places to hunt seals. . . . They said when there was a delay in the ocean being open for hunting, those poor people [lacked food]; my mother was a girl and was old enough to gather greens, she said that they would go here to Kangirrluar [Toksook Bay], probably down there [on the shore] to fish with hooks and lures all day. And they didn't catch a lot.
> And then she said that they would go up there, using ice picks as their tools, and gather *ariqat* from above Up'nerkillermiut. . . . She said that people would also have their children eat those, having them survive off of those during the famine as a supplement. . . .
> She said that the small children would eat those *ariqat*.

Nunivak elders (Nuniwarmiut Taqnelluit 2018:126) describe the same plant, but there, too, the species could not be identified. Martina John (May 2019:227) commented wisely on the regional variation in plant knowledge: "People are amazed when they go down [to Nunivak and see people eating] those. . . . There are probably those on the shore here, but people don't know them."

Looking west toward Up'nerkillermiut along the north shore of Toksook Bay, where Lizzie Chimiugak and Albertina Dull gathered plants when they were younger. ANN FIENUP-RIORDAN

Simeon Agnus discussing plant use during a trip around Nelson Island, July 2007. Theresa Abraham is holding a bunch of wormwood, which she picked close by. ANN FIENUP-RIORDAN

NAUCETAAT IINRUKTUKNGAIT
MEDICINAL PLANTS

✳

Yungcautnguuq nunam qainga tamarmi
The entire surface of the land is medicine

Yup'ik food generally continues to be viewed as healthful. Marie Alexie (June 2016:234) declared: "Since Yup'ik foods are real food, they are a help to the body." Plants are seen as particularly beneficial. Joshua Phillip (December 1985:128) of Tuluksak described how plants are medicine to the body:

> Our elders used to say back then, those with TB [tuberculosis] got feeble in winter, they would be weak by the end of winter. They would get thin. But in spring, beginning when the plants began to sprout and they began to eat the plants all summer, they would get alert and strong again.
>
> They would get their strength. Plants that grow on the land are medicine to the body.

Grace Parks (June 2016:240) declared: "*Nunam qainga iinruuguq* [The top of the land is covered in medicine]. And it's covered with food, too." Elizabeth Andrew (April 2008:211, 221) agreed: "There are many kinds of medicine, the medicine of the land. I learned those through our ancestors. . . . Everything on the land is medicine."

Barbara Joe (April 2012:61-66) of Alakanuk spoke at length on plants as both food and medicine:

> They would gather plants in summer. Plants were their foods. In the spring, we'd start eating *iitaat* [edible lower stems of tall cottongrass]. Then when plants grew, they ate those plants until they got hard.
>
> And these *quagcit* [sour dock], they cut them into pieces and made them into extremely tasty *akutaq*. They are very delicious. Those plants prevented us from being ill.
>
> Also, when they would open up salmonberries in winter, when we'd have colds, they had us drink berry juice in the morning. And not filling the cup full, they let us drink a small amount of berry juice. They say those are medicine. Since those things were medicine to their bodies, today I've come to understand what they're like; they say eating those berries makes one strong since they have many vitamins.
>
> Since we no longer eat those things we were raised on, we poor things today tend to get weak. . . .
>
> Those things they raised us on, some of us happened to catch those things. They say their foods prevented them from becoming too ill. They continually ate their foods and became elderly.

The focus of traditional Yup'ik concepts relating to sickness was on prevention rather than cure. "Keeping well" was an underlying motive behind the majority of *alerquutet* (admonitions) and *inerquutet* (proscriptions), with myriad rules prescribing how a person should live so as not to get sick. Many plants were used medicinally to treat relatively minor ailments such as headache, earache, cuts, and abrasions. As anthropologist Margaret Lantis (1959:54) aptly pointed out, the strength of the Yup'ik system for disease control lay in keeping healthy. Use of medicinal plants played a large role in this process.

Caiggluut wall' qanganaruat
Wormwood

Perhaps the most widely used medicinal plant in southwest Alaska, worm-wood (*Artemisia tilesii*) or "stinkweed" (because of its pungent aroma) is known by several names: *caiggluut* on Nelson Island and along the coast as far north as Norton Sound, *qanganaruat* (lit., "imitation squirrels") along the Kuskokwim, and *naunrallraat* and *naunerrluut* (lit., "bad plants") along the lower coast, including Quinhagak and Platinum. Paul John (September 2003:227) shared his understanding of why different names for wormwood were used in different parts of the region:

> In my village [Toksook Bay] wormwood is called *caiggluk*. And down at Kipnuk, people also call them *caiggluk* since there is someone named Qanganaq [Squirrel] there. And since there's someone named Caiggluk at Eek, [wormwood] is called *naunerrluk*. At Quinhagak some also call it *naunerrluk* because there is someone named Caiggluk there. And around

Caiggluut (wormwood), also known as *qanganaruat* and *naunrallraat*, is among the most important medicinal plants in southwest Alaska. KEVIN JERNIGAN

Wormwood. KEVIN JERNIGAN

here [in the Kuskokwim River area], this [plant] is probably called *qang-anaruaq* since there is no one named Qanganaq here. I have heard them being called those names.

Flowering and seed-bearing stalks of wormwood are considered male, while those without flowers are considered female. People disagree on whether or not both male and female parts of the plant are effective medicine. Francis Charlie (January 2013:362) recalled:

> Those *caiggluut*, since we always have a supply, one day I took a bag to use, and as I was putting my shoes on, my wife asked me, "What are you about to do?" I told her that I was going to gather some *caiggluut*. She got closer to me and told me not to gather their males.

Theresa Abraham holding up a stalk of dried wormwood during a plant gathering in Anchorage, February 2019. ANN FIENUP-RIORDAN

She said one of the people of Mountain Village told her that, like Labrador tea, not to gather their males. Then I asked her, "Why?" She said their males are dangerous. [*laughter*]

Although she told me that, I included the leaves of the male plant. As you know, the female plants are all leaves. Since they are all the same, I picked them.

Neva Rivers (May 2004:331) agreed that all parts of *caiggluut* are used as medicine. Elizabeth Andrew (April 2008:206) noted that even the roots are edible: "You pull their roots out and wash them. And after washing them, if you chew on them, your mouth will be numb."

People harvest wormwood during different times of the year. Some say fall is the best time to harvest the plant. Others pick and eat the new shoots in spring, chewing on them and swallowing their saliva. Theresa Moses (September 2003:229) noted that in Chefornak, children were given the tips of "male" wormwood plants to eat: "We would chew them and swallow them." According to James Guy (June 1995:13) of Kwethluk, "Eat wormwood when they first sprout so that you can be physically active. They are medicine before they mature." Theresa Abraham (February 2019:41, 43) described how her mother harvested wormwood tops in early summer, spread them outside on a piece of plywood, and let them dry: "She said when these [wormwood] are at this stage their medicine is strong. . . . They say taking their tops [as medicine] is more powerful than taking their lower stems." Marie Myers (September 2003:228) of Pilot Station noted that dead plants can also be harvested: "In spring they cover them with hot water and cook them. Even though they are dead plants, they are useful." Nick Andrew (February 2019:67) agreed: "It's okay to pick them even in winter. . . . While they're still standing outside where they grew, withered, they say they still work."

Worldwide, there are a number of species of *Artemisia*. The one found in southwest Alaska, *Artemisia tilesii*, contains the terpene compound isothujone, which has codeine-like properties binding to opiate receptors in the brain and dulling pain (Overfield et al. 1980:97).[4] In part because of this therapeutic effect, wormwood has a wide variety of uses. Martha Mark (January 2011:280) declared: "Wormwood is medicine for everything." Many people either boil wormwood in water or simply cover it with hot water, then drink the liquid as a healthful tonic. Carrie Pleasant (June 1995:26) of Quinhagak noted: "Some people boil the water first, then pour it over the dried wormwood like making tea. But I cook them by boiling them." Annie Andrew (July 2000:22) of Kwethluk said that she dried the plant, then poured water on it and drank it like tea:

> After feeling sickly with a sore throat, these healed me after I started to sweat. . . .
>
> There are many ways that you can use this for healing your body. When you have bronchitis or pneumonia or stomach trouble or if you get sores in your mouth, or if you get a sore throat, or cancer, any sickness that really bothers you, you can cook this when it's dried. Just take a batch and put it in a little pot and cook it, but some people boil it. Whenever I use it, I don't boil it. I keep on stirring it until it nearly boils and take the heat out. After it's cooled, just take a small amount every day. This will take away your pain or sickness, it will relieve it or even heal it, especially the colds that bother you.

Wormwood. SHARON BIRZER

Theresa Moses (September 2003:226) also prepared *caiggluk* for drinking:

> Some used them to heal their bodies and drank them in the morning.
> They didn't eat it, but drank it. Some used it like medicine at noon and in
> the evening to heal their bodies and their sicknesses.
>
> I make it for drinking. I use hot water and then put them in it and let it
> boil covered and leave it alone for a while. When I remove it, it will have
> already gotten strong.

Paul John added: "They are good as tea. I like the taste better than Labrador
tea when we drink tea."

Mike Andrews (August 2011:646) also prepared wormwood as a drink: "I
cook them in a pot. And I boil them for a short while, and when they cool
off, I pour them inside a bottle, but not these [leaves or stems]. I drink a little periodically from that bottle of juice, even at night. These *caiggluut* were
medicine used by people long ago." Martha Mark (January 2011:279) noted:
"When my grandchildren have sore throats, I have them drink a small
amount with a small juice scooper. They get better right away. We'd wake the
next day, and their sore throat would be better."

Mike Andrews (August 2011:646) also testified to the healing powers of
wormwood juice: "A person who had cancer, a doctor evidently sent him
home and said that he would die. And when that person went home, they
say he drank the juice that seeped out of these [wormwood plants]. They say
eventually that person started eating, and they say he recovered. They say
these saved him." Finally, Nick Charles (December 1985:141) described cooking *caiggluk* and drinking the juice to cure his tuberculosis. Joshua Phillip
added that he drank *caiggluk* juice to treat heartburn and indigestion, and
Paul John quipped, "They are very good Eskimo Tums."

People also chewed on fresh or dried wormwood. John Andrew (April
2019:258) recalled: "They say they peel and eat them when they are newly
grown [young shoots], before they taste sour. They are weak when they are
new shoots. When they start to get big and grow tops, they are very powerful."

Mike Andrews (August 2011:646) continued: "Their tops are good, chewing on them and swallowing their juice." Theresa Abraham (July 2007:625)
recalled: "When I feel like I'm about to get a cold, I chew on those tips [of
wormwood]. When their juice collects in my mouth, I swallow it, and when
I get tired of chewing on them, I take them out. When I do that, I later find
that the cold that I felt coming on has gone away."

Another common use of wormwood was to treat cuts and sores. According to James Guy (June 1995:13): "Wormwood is good on cuts. When a cut is too large and gets infected, you can use it to treat it. After you soak [wormwood] to get it soft, make it pliant and then place it on the cut and wrap it with a bandage." Neva Rivers (May 2004:331) noted that one could use either fresh or dried wormwood: "Even when they dry up, we can soak them or just chew them, and when they soften, put them on sores. If we tape it on the affected area, we leave it on overnight. The next day, I will remove [the tape] and [the wormwood] will have stuck to it. The dried peel would be on the cut that had pus in it, and it would be in the process of healing."

Wassillie B. Evan (July 2000:22) said that he had used wormwood to heal a cut by mixing the leaves with seal oil and using a tobacco leaf to cover it. He found his cut clean the next day. Theresa Moses (September 2003:225) reported softening wormwood leaves by rubbing them in a circular motion and using the resulting cotton-like material as a bandage to cover cuts as well as a treatment for canker sores.

A similar technique was used to treat boils. Annie Andrew (July 2000:22) described the process:

> If it is used on a boil, remove the leaves, rub them [between your palms], and dry them. After they are dried, they are easier to use. Then use a little bit of shortening after it gets a little numb, add a little bit of sugar, then cover the boil. This can heal the pain that is inside the boil and let the infection out. Then we can put another cover on it with another leaf.
>
> I changed the leaves twice in one day, but I kept one on the whole night and the boil started to itch, then the infection from the boil came out. After that, the pain was gone.

Breathing in the steam from wormwood is also said to reduce congestion and promote healing. Joshua Phillip (December 1985:138) noted:

> He can cook wormwood and burn rocks until they turn red hot. Then put [the rocks] in the water [containing the wormwood] and put a chair nearby, and sit and cover himself and steam. He would bend his head down and take in the steam. When it cooled off and he finished the steam, he would find that his body was covered with thick slime. He would peel it off and let it fall. The vapor jelled on his body. He would remove that and wash up. His sickness would come out, that would pull it out.

To this day, people regularly use wormwood in steam baths to promote sweating. Timothy Myers (September 2003:228) of Pilot Station recalled: "In my village since they started taking steam baths, they use these when they bathe. They put them in the container of water that is used to pour on the rocks and let it boil, then pour the liquid onto the rocks. It is very good to use, and it seems like there is a lot more sweating." Paul John (September 2003:229) added:

> When my wife and I take a steam bath, she sometimes pours hot water in the basin after putting [*caiggluk*] in it, and she pours the water [with *caiggluk*] on the rocks.
>
> They also swat themselves on their back or body with them [during steam baths]. They have many uses. Since some people believe that they work, it helps them.

John Andrew (April 2019:254) noted that people in Kwethluk also use *caiggluk* to swat their bodies during steam baths: "Those with sore bodies and tight muscles, and even when they have muscle spasms, after dipping [*caiggluk*] in hot water in the steam bath, they use it to swat the area that is throbbing with pain." Theresa Moses (September 2003:229) described the smell of wormwood when added to water in the steam bath: "When they pour the water, it smells just like Vicks. It makes the breathing deeper."

Annie Cleveland (January 2011:282) of Quinhagak emphasized the use of wormwood to relieve congestion:

> If you have a cough and you can't seem to get the congestion out of your chest, after steaming with that in the steam bath or if you just cover yourself with something and place the bowl here [below your head], the next morning, sometimes by itself, when clearing your throat, it suddenly comes out. It's very good to take congestion out of the chest. It's much better than Western medicines.

Theresa Abraham (July 2007:723) had used wormwood in the steam bath as a hot compress to successfully relieve her pain:

> I braided [wormwood plants] together, and inside the steambath house where it's hot, I placed them along one side of my body, and covered it and held it . . . with my arm. At the time, I was short of breath on this side. But I'd do that in the heat, placing those [*caiggluut*] on there. That

injury that had been painful, after covering a large area, slowly got smaller, and eventually the pain . . . completely disappeared.

Many reported using wormwood as a poultice. Anna Agnus (July 2007:620) explained:

I just placed them on my body and kept them there. . . . From that point on, because of that particular plant, my body healed after I suffered from constant back pain. They indeed work when they are placed directly on the body, . . . sleeping with them on my body and walking with them on my body, too.

One day, when that person there [her husband, Simeon Agnus] was suffering from pain when he was working, although he didn't seem to believe in their ability to heal, I braided them together, and after dipping them in hot water, I placed them on the part of his body that was in pain. This person got better, and he finally believed in their ability to heal.

Simeon elaborated: "I covered [the *caiggluk*], and it seemed that when I slept on top of [the painful area], . . . the heat started to reach the inner part of my body; it was making a poking sensation. My, when I woke up, one side of my body where I was in pain was good. That type of medicinal plant caused the pain to go away, and I healed."

Francis Charlie (January 2013:368) gave another example of using wormwood as a poultice to heal an open wound:

A man evidently told about how *caiggluk* saved him. He said he had something that wouldn't heal on his leg. One of the men who saw it told him to cook [*caiggluk*] at a heavy boil, and when it cooled down, to wash it thoroughly. And when it no longer had any debris on it, he should immediately place a number of [leaves] as they were on that thing that didn't heal.

Then they say he bandaged it with a piece of cloth that wasn't too airtight. He kept it there. The next day, he opened it and saw that the *caiggluk* was stuck on it. Then he replaced it with another. He said that thing that wouldn't heal, starting from its outside got smaller and smaller. He said *caiggluk* healed it. He only stopped treating it when it healed.

It should come as no surprise that such a potent plant was used to promote good health generally. Ruth Jimmie (July 2007:622) noted: "From what

I know, Acac'aq used to braid some *caiggluut* and always wore that as a belt. She always had a supply of those; she'd use them in that way as a belt around her waist during the day." Just as wormwood could be used as a poultice to promote physical healing, it could also be rubbed over a person's body to purify that person following illness or death of a family member. Finally, the smoke of burning wormwood, like *ayuq* smoke, was used as purification. Theresa Moses (September 2003:230) recalled:

They would use these for *tarvaq* [cleansing oneself with smoke]. The tips would be smoking. Then they would also inhale [the smoke].

And when they put it inside and underneath their garment, they would open this [neck opening] and place [the smoking plant] down there [near the garment's hem]. They would stand up and let the smoke go out through [the neck opening], and they called it *qumigturluni* [putting something inside]. The smoke went inside between one's clothing and one's body. . . .

They wanted [their sickness] out and said that this kind [of plant] took out [the sickness]. They burned it, and they smoked their sickness out with this. . . . That was what they did when I watched them.

And once when I watched those who were going seal hunting, I was amazed to see someone light something on fire over there and then go. Then he placed his kayak on the kayak sled, and he quickly went over that fire and left. I then asked my late older sister, "Why is he doing that?" She told me that he was cleansing his kayak before going seal hunting.

These were medicine from way back then. They were smoke cleansers. They would also put [the smoke] inside their clothes. They let [the smoke] come out [through the neck opening]. And it seems like it was strong.

Paul John ended our discussion by singing *"Tarvarnauramken"* ("Let me purify you with smoke"), a well-known Nelson Island blessing song, to dramatically portray the process of purification.

Ayuq
Labrador tea

Ayuq (Labrador tea, *Rhododendron tomentosum*, formerly *Ledum palustre*) is found in tundra areas all across southwest Alaska. *Ayuq* is widely used, both made into a healthful drink and as medicine. Its name derives from the verb base *ayu-* (to spread, to go farther and farther away) and may refer to the plant's ability to carry away illness, as in the expression "*Ayum ayuurutaa* [Labrador tea takes it away]."

Theresa Moses (September 2003:224) said that in the past when people ran out of tea, they would make tea using *ayuq*. Martha Mark (January 2011:284) recalled that her grandmother always had a large supply inside a grass bag, and when he was young, Mike Andrews (December 2011:233) remembered seeing *ayuq* hanging along the wall of his family's home. Raphael Jimmy (January 2013:351) noted:

> When I became aware of life, when I observed them, they used it as medicine and even as tea. Since my parents constantly ran out of supplies in the fall, my mother would see the amount of Lipton tea, and if it seemed like they would run out, they would pick [*ayuq*] and add to it and mix it, and the amount suddenly increased. They didn't run out. And when I came to observe things, my mother was never without a supply.

Joshua Cleveland (January 2011:281) agreed: "We also like to gather *ayuq* and add those to tea. . . . We've gotten to where we like to have a constant supply. We don't like to be without those. And when they finish, we tend to look for them."

Although *ayuq* could be gathered throughout the year, many said that *ayuq* should not be gathered when it is in bloom. Cecelia Andrews (April 2019:227) explained: "I don't pick them with flowers on them. . . . I only saw people picking them before they grew flowers." John Andrew (April 2019:227) added: "In the summer when they have flowered, they aren't strong. But in the winter after they had wilted, when the plant has grown, one that is like that is stronger."

Ayuq (Labrador tea) in bloom. KEVIN JERNIGAN

As with a number of species, people distinguished between male and female plants: *angucaluut* (male plants) had flowers and seeds (known as *nasqut* or "heads") while *arnacaluut* (female plants) did not. Barbara Joe (April 2012:65) noted that her parents had her chew on the "heads" of male *ayuq* plants when traveling on the tundra: "They say the tops of *ayut* are also medicine." Wassillie B. Evan (July 2000:27) commented: "After you chew on the tops of *ayuq*, there is an aftertaste. It tastes like Vicks."

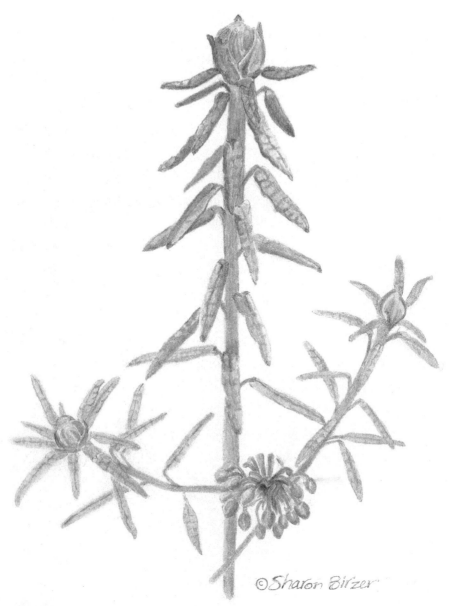

Labrador tea. SHARON BIRZER

Paul John (October 2010:223) recalled the important admonishment to chew on *ayuq* to help quench one's thirst and avoid lethargy when traveling in the wilderness.

If [a person] happened to walk a good distance away when conditions were safe, back when they tried to catch fur animals on foot as a source of income, if he became thirsty and he didn't have a tool to make a hole in the ice, as much as they warned people against eating snow, they'd tell a person that if he saw some Labrador tea, he should take some. They say chewing on Labrador tea and continually swallowing the saliva that forms prevents one from becoming weak and lethargic.

But they say if he doesn't do that and continually eats snow, he will start to become weak and lethargic, and he can start to have problems walking. I also experienced that, the condition where one continually falls although there's nothing to trip on.

Cecelia Andrews (April 2019:224) always had her children eat an *ayuq* leaf when they returned home after a long trip:

If someone has been gone from his home for a long time, when they finally arrive home, because their environment is different, something happens to them, they get sick. For that reason, you have to eat a little bit of soil when you arrive. That's what they used to tell us to do. Because these [*ayut*] are [of the land], I give it to them, even if it isn't soil. . . . They say a person who hasn't done that won't be in good health.

Labrador tea growing near Scammon Bay,
August 12, 2017. JACQUELINE CLEVELAND

Many elders testified to the medicinal value of *ayuq*. Martha Mark (January 2011:289) noted: "Labrador tea has been medicine since long ago; when I became aware of life, I saw that they were medicine." *Ayuq* leaves and stems can be added to tea for good health. They can also be brewed alone as a healthful tonic for treating coughs and colds. Joshua Cleveland (January 2011:281) noted: "Some people who have colds evidently also take them as medicine besides adding them to tea. Constantly consuming that seems to help us. Although others are having colds sometimes, it's as though we don't have colds when we consume those." George Pleasant (January 2011:289) described his experience:

> One time, after looking around for cough syrup or cough drops, since there were none, I headed downriver from the village and gathered some Labrador tea. When I arrived home, I placed them inside a kettle and poured water and cooked them.
>
> After cooking them for a while, I poured [the liquid] inside a jar and cooled it. When it was cool, I filled that scoop [from a Tang juice container], and although I didn't drink a lot, I drank some once in a while, timing it like taking regular medicine, every four hours. My cold improved; I no longer had a cold.

Theresa Abraham (July 2007:627) described how her younger brother, John Avugiak, suffered from continuous diarrhea when he was young. At that time there was no clinic in their village: "After trying different medicines, my mother went up to the land behind our village and shoveled the snow with a wooden ladle and picked some [*ayuq*]. Then she boiled some for him. From that time on, she'd give him *ayuq* to drink. His condition started to improve right away, and his diarrhea went away. I think if that *ayuq* hadn't healed him, he would have died."

Elders warned that one is not to take a large amount of *ayuq*. Theresa Abraham (February 2019:55) said: "They say it's not good when people consume too much." Cecelia Andrews (April 2019:223) agreed: "They tell people not to take a large amount." John added: "But they said one must take a small amount from time to time. . . . They are very good when added to tea."

John Walter (July 2007:629) used *ayuq* like *caiggluk* in the steam bath to clear his sinuses: "Before we spill water over the stove, we place [*ayuq*] on [the stove] and have it give off smoke. Afterward, while we're taking a bath, they let these give off smoke. When we are done, [our sinuses] aren't stuffed, it clears the inside of our noses. They have started to use those in that way."

Ayuq was central to traditional healing practices. Elsie Tommy (June 1992:28) described a woman who had acquired healing hands when she found furry caterpillars in a vole cache, then laid her hands over the insects and allowed them to penetrate her body (Fienup-Riordan and Rearden 2017:334–37). Later, Elsie observed the woman placing her hands on a patient's body to heal him, and then using two *ayuq* plants to help her remove her hands: when she placed her hands on her patient, she created a nonmaterial connection that the *ayuq* plant helped to disengage:

> When they placed the two *ayuq* plants there, after a moment, she moved [her hands] this way over them and detached them. . . .
>
> They say since those furry caterpillars stay on [*ayuq*], [the plants] helped them in their work for prying off [her hands].
>
> They still see them sometimes these days. [The vole cache] would be filled with all furry caterpillars. They would be lined up. They looked like they were boiling.

Like *caiggluk*, *ayuq* leaves and stems were burned as an act of purification. James Guy (June 1995:21) recalled: "Our ancestors used it to purify themselves, as a way to cleanse themselves and become active again. When they did that, they burned a bundle of *ayuq* and doused their bodies with the smoke." Martha Mark (January 2011:284) remembered: "My grandmother used to do that to me. She'd light *ayuq* and then she would [brush it over me] and say, 'Let me *tarvaq* [purify] her.'" Barbara Joe (April 2012:65) had experienced the same thing: "When we young ones would take baths, after taking baths, they would burn [dried *ayut*] and shake them over our bodies. They say they are *essuircautet* [those that are used to cleanse and purify], they are good. They say when we [cleanse] our bodies with *ayut*, we lose our impurities."

George Pleasant (January 2011:284) noted that homes as well as their occupants could be purified in this way: "They'd light the *ayuq*. I used to see people do that once in a while long ago. And sometimes my grandmother would light it and have it burn and spread [the smoke] around the house. She would purify it with *ayuq*." Annie Cleveland (January 2011:284) described how *ayuq* smoke was spread around a home following a death: "When they'd take a deceased person out of the house, they'd [purify] the home with *ayuq*. Did they do it to freshen the smell of the air or to take the spirit out? I don't know the real meaning of that, but that's what they used to do." Mike Andrews (December 2011:234) had observed the same thing: "When my father finally heard about someone dying, even though that person was from far away, they'd cry a little and then light those *ayut* during the time they heard [about the death], and they'd [purify] the inside of the home." Raphael Jimmy (January 2013:356) remembered:

> When I came to observe things, after putting the deceased in a seated position, then my mother would light a small amount of *ayuq*. Starting from the corner, when they took [the deceased] out of the home, she would spread *ayuq* smoke in the area, following the one who was going out. And she would also put *ayuq* smoke on the small shelves, thinking in her mind, "Sickness, illness just left following that person who is dead." Then when she went outside, she didn't bring [the *ayuq*] inside but discarded it. That's how my mother used to carry it out.

Lawrence Edmund (December 2011:231) of Alakanuk described using *ayuq* to purify hunting equipment: "Purifying with *ayuq* smoke has a good consequence." Jacob Black (March 2016:145) of Napakiak had boiled his hooks using *ayuq* to make them more effective:

> The elders said that when they were unlucky at catching using their hunting and fishing implements, after getting *ayut*, they would boil them. Those were what they used to make their implements better at catching.
>
> I tested it out with my fish hooks. After removing my fish hooks, I boiled some *ayuq* and used its water. Then after [the water] got dark, I dipped [the hooks] inside there. It was very true.

David Martin (May 2004:100) recalled the use of *ayuq* smoke to make a hunter appear bright to ocean animals:

> Following the custom of his ancestors, one who was going down to the ocean for the first time would light a fire, and after filling it with *ayuq*, after removing his belt, he would place heavy smoke all over his body. Then after doing that, he would take his kayak and go through the fire and go to the ocean. He would have his kayak go through it, too. They say if he followed the tradition that was passed on from his ancestors, he would appear bright to the persons of the ocean. Since our elders explained everything, they said one who was going down to the ocean for the first time is brightening himself so that the persons of the ocean wouldn't be offended by his appearance.

Alakanuk elder Joe Phillip (December 2013:367) noted that everyone who traveled to the ocean purified their bodies as well as their equipment with *ayuq* smoke: "When they are about to travel down [to the ocean], as you know, some women like to come along when traveling. They said that after burning *ayuq*, they would purify all of them, as well as their clothing and things. They said after doing that they would travel down to the ocean." John Andrew (April 2019:225) said that the same was true for those traveling in the mountains:

> Up in our village [Kwethluk], those who travel to the wilderness, when some of them had been sick, when they were going to hunt for the first time, they were told to smudge themselves with these *ayut*. They would gather a large amount, and they would smudge their clothing and traveling gear. They said they were trying to remove their impurities. They say those who haven't done that are inaccurate or unlucky when they go for the first time after they had been sick.

Francis Charlie (January 2013:351, 355) offered a slightly different explanation for the effect of *ayuq* smoke but with the same desired results: successful hunting.

> A person who cannot see or cannot catch although he hunts, they have him purify himself with *ayuq* smoke. . . .
> Some people used to say, "*Ayum ayuurutaa* [Labrador tea takes it away]." They say its smoke was like a shield to animals they hunted. Its smoke was like a barrier. If I cover my face, I cannot see this person although he's close. They say that's what they're like. . . . When a person purifies himself with *ayuq* smoke, it's like [the animals] cannot see him.

Nastasia Larson (March 2017:185) of Napaskiak gave a detailed account of traveling to spring camp in the lake country south of her village just after she had menstruated for the first time. When her family reached the Luumarvik River, she was told to gather *ayuq* plants and put them in her pocket. When they reached their spring camp, the man who was leading them lit the *ayuq* and smudged them. Nick Andrew (February 2019:53) noted that in this case the *ayuq* smoke functioned as protection: "My mother would do that to her brother's daughter when she was about to go to the tundra. . . . They say that these things, including *ircenrraat* [other-than-human persons], things that are unseen, they throw things at a person who isn't smudged. That person's body suddenly swells." *Ayuq* smoke, they say, can repel unwanted spirits.

Ayuq smoke is still used today in public and private purification ceremonies. Joe Phillip (December 2013:367) noted that when he was training to be a deacon in the Catholic Church, elders told the priests about their use of *ayuq*, and it was subsequently incorporated into the Mass. *Ayuq* has since been used to perform blessings in other public and private events. Mike Andrews (December 2011:231) recalled:

> Quite a while ago, I attended a potlatch at St. Mary's from Emmonak.
>
> Then the next day [after they danced], they brought the gifts they would distribute inside the hall, the things they would distribute to their guests, including the people of Pilot Station, who were gathered together down there.
>
> Then the priest, holding those *ayut, tarvat* [plants burned as purification], he put smoke on them using a bird's feather; before they distributed those things, he put smoke on [the gifts] and circled around them in this [clockwise] direction. The priest also blessed them with Holy Water. I watched people doing that at St. Mary's. They carried out old traditions.
>
> When they were done, they finally distributed those things. . . . That's the first time I saw that at St. Mary's; it was quite a while ago. . . .
>
> You know how the Catholic Church has incense; they say Yup'ik teachings and the church's teachings are the same. They compare *ayuq* to incense in the church.

Raphael Jimmy (January 2013:357) noted: "These days, the priests still carry that out. They cannot be without it. They use it to cleanse, and when there are many people at church, they cleanse them and have the smoke take [impurities] away." Finally, Raphael noted that only those who believe in the power of *ayuq* can receive its benefits: "One has to receive it well. If something is wrong with you or if you're sick, if a priest or deacon uses *ayuq* to bless you, then you would accept and receive it. Through the smoke, your sickness will leave and you will gradually get better. Only if you believe in it." Francis Charlie agreed: "You have to accept it and receive it as being real."

Belief is equally important for *ayuq* to be effective as a medicine. Raphael (January 2013:354) concluded: "If you eat this [*ayuq*], or if you eat the top of it when you have a sore throat, when you take it, you have to let it work in your mind. You must believe in it. They say it will only work then. But if you just eat it like eating dry fish, it is nothing."

Atsaruat (wild chamomile). KEVIN JERNIGAN

Atsaruat wall' itemkeciyaat
Wild chamomile

Wild chamomile (*Matricaria matricarioides*), also known as pineapple weed, is a low, green plant with small, ball-like yellow flowers. Wild chamomile is widely known in southwest Alaska as *atsaruat* ("pretend *atsat* [berries]"). Other names include *itemkeciyaat* (from *itemkar-*, "to kick lightly") and *kitengkaciyaaret* (from *kitengpag-*, "to kick hard") on Nelson Island, and, in Kotlik, *itegmik* or *itegmigcetaat* (from *itek*, "toe piece of a boot"), perhaps referring to the light sound the ripe flowers make when kicked. *Atsaruat* are widespread in southwest Alaska, growing in open areas where people walk or where the ground has been otherwise disturbed. Martina John (November 2007:319) accurately observed that they are abundant in places

Wild chamomile. JACQUELINE CLEVELAND

where dogs once stayed: "They call them *itemkeciyaat*. When we'd arrive in Nightmute [when returning from fish camp in July], they were abundant." Alice Rearden (November 2007:320) of Napakiak commented that she used to eat their yellow tops when playing outdoors as a child, and Martina added: "When we were playing, we'd pretend to make them into *akutaq*." Cecelia Andrews (April 2019:229) remarked: "We, too, pretended that they were berries. When we were playing outside, we'd take these and eat them." The name *atsaruat* ("pretend berries") fits both their appearance and their use.

Denis Shelden (January 2013:344) of Alakanuk said that their leaves and flowers were good added to soup. John Andrew (April 2019:227) described their use along the Kuskokwim: "When they start to grow upriver but before they grow tops [flowers], some people pick them, cook them, and eat them. Or even if they look like this [with flowers], after boiling them, they prepare them into a solution to drink for those who are sick. They say they cause a lot of sweating."

Theresa Abraham (April 2019:230) said that her mother harvested and dried *atsaruat* for medicinal use: "She used to gather these. . . . Then she would braid them and hang them outside to dry, then store them somewhere

Barbara Uisok of Kotlik holding wild chamomile near her home, July 11, 2017. JACQUELINE CLEVELAND

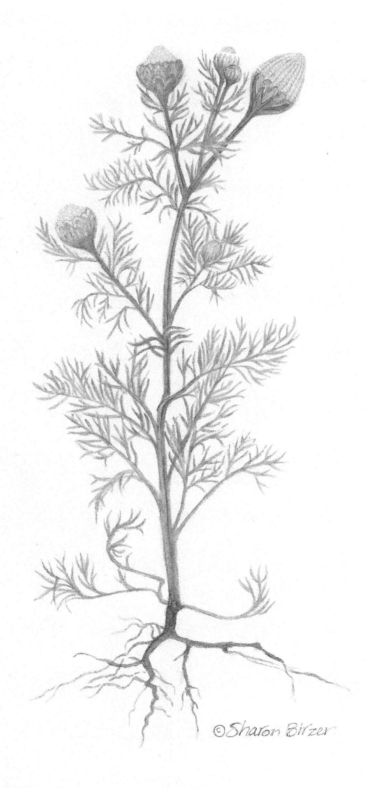

Wild chamomile. SHARON BIRZER

when they dried." Ruth Jimmie added that her mother had also braided and dried *atsaruat* and used them as a belt to ward off illness. Joe Phillip (December 2013:365) also commented on the use of *atsaruat* as medicine: "They used to eat them when they would cough and spit blood from tuberculosis. They say those make one recover." The flowers can also be boiled, and the liquid taken either cold or as a hot, soothing beverage. Placid Joseph (December 2013:365) of Alakanuk noted: "When one has a cold, they say they are good when cooking them [and drinking] the juice."

James Guy (June 1995:13) noted the use of wild chamomile to help sweat illness out of the body: "After you have cooked some, put them along with their juice in a basin. Then if your knees are bothering you, place them over the basin and cover them up with a towel and steam them. When they start to sweat, wipe the sweat off. Whatever is bothering you will come out. The sweat will be in the form of slime; when that comes out, the arthritis ailing you will be healed." Joshua Phillip (December 1985:138) agreed that, as with wormwood, when one finishes such a steam, one's body would be covered with slime, and removing this slime would pull the sickness out of one's body.

Finally, Peter Jacobs (October 2003:59) described using *atsaruat* to predict the salmonberry harvest: "When the wild chamomiles are large, they say that the salmonberries are ripe." Placid Joseph (December 2013:365) agreed: "They say they ripen at the same time as salmonberries. When those [*atsaruat*] grow, they use them as indicators. They would say it's also time to pick salmonberries."

Tumaglit wall' kavirlit
Low bush cranberries

Tumaglit (low bush cranberries, from *tumag-*, "to be bitter tasting," *Vaccinium vitis-idaea*), also known as *kavirlit* (lit., "red ones"), are widely recognized as having medicinal value. Martha Mark (January 2011:286) declared: "Low bush cranberries have always been medicine since long ago. And when someone has a cold and his throat has a tendency to be dry, then you can cook low bush cranberries and have him drink a little bit of their juice or have him eat low bush cranberries. My grandmother also used to do that when her children's children had colds." Martha also described using the leaves of *kavirlit*:

> The leaves of the small *kavirliq* bush are medicine. When my blood
> pressure starts to rise, he goes and gets me that [type of leaf] when he

travels. Then I pour hot water on it and prepare a liquid remedy. I drink it in the morning and at night. My blood pressure goes down. I've gotten used to taking that kind of medicine.

There are many *kavirlit* outside the church down there also in my village and down below our village along the shore. Many people use them as medicine by cooking them. They are good made into a liquid remedy.

Low bush cranberry stems are also said to be good for asthma. Elizabeth Andrew (April 2008:212) noted: "Those with asthma prepare a potion to drink out of those [low bush cranberry stems]. And we also drink those, even though we don't have asthma." Martina John (July 2007:642) added: "They have mentioned that low bush cranberry stems are extremely good as medicine for asthma. Because they mentioned them, I've been picking them recently so that I can use them. They prepare them by cooking them and making juice out of them and using them in that way. I'm grateful [to learn about their health benefits]."

Ruth Jimmie (April 2019:22) recalled picking and eating last year's *tumaglit* in spring: "They were very sour, but they were medicine. . . . People constantly took medicine as they lived." John Andrew noted that upriver along the Kuskokwim, people gathered *tumaglit* when it started to get cold: "When those *tumaglit* form juice, they can use them as medicine. They'd let them use them as eye drops for those with snow blindness. They'd get their eyesight back right away. They'd let me use them as well." Theresa Abraham (April 2019:99) had also been treated with *tumaglit* juice for snow blindness:

In spring, when it became bright outside, some people's eyes stung badly. That happened to me. Then my mother wanted to drip the juice of a *tumagliq* [low bush cranberry] into my eye. She told me to open my eyes and to stay still, and my eye stung very badly.

I also saw another person do that. They just dripped *tumagliq* juice in the eyes of someone who became snow blind. . . . It stung and then got better.

◄ *Tumaglit* (low bush cranberries). KEVIN JERNIGAN

Melquruat wall' maqaruaruat
Cottongrass

Melquruat (lit., "pretend *melquq* [fur]") is the Yup'ik term for both white cottongrass (*Eriophorum scheuchzeri*) and Russet cottongrass (*Eriophorum russeolum*), also known as *maqaruaruat* (lit., "pretend *maqaruat* [snowshoe hares]") and *ukaviruat* or *ukayiruat* (also "pretend snowshoe hares," from *ukayiq*, cognate to Siberian Yupik *ukaziq*, "snowshoe hare"). Cottongrass has several medicinal uses. Katie Jenkins (June 2016:235) of Nunapitchuk described using them as a poultice to treat cuts and boils: "After soaking them in seal oil, they are good on boils. And when a cut worsens, after soaking it in seal oil, after putting it on a cut, then they bandage it." Marie Alexie added detail: "My grandchild picks some when they grow. Then she puts them away. Then when she gets a cut, after rubbing it with shortening, she covers and bandages it. She'd open it up, and we would see that it had already healed." John Andrew (April 2019:68, 265) agreed:

> They say those who have boils, the [boil] has a *yua* [root of the boil, lit., "its person"]. They say after covering it with these [*melquruat*] that had been soaked with seal oil, they keep it covered. They say it makes its hole enlarge, so that the pus can leak out better. . . .
>
> They used it like cotton. And they said injuries that couldn't heal right away, after covering it with that, and after putting a layer of wormwood on top of it, they then bound it [and it would heal].

Martha Mark (January 2011:280) noted that *melquruat* grow in abundance when there are going to be lots of salmonberries. Marie Alexie (June 2016:235) agreed: "They said they are indicators for berries. When [white cottongrass] grew [in abundance], they would be happy . . . and say that lots of berries are going to grow." Also when the cotton begins to blow in the wind, they say that berries are ripe. Grace Parks added: "You probably see them when you pick berries, covered by cotton. They said those keep those salmonberries warm." Grace's mother, Katie Jenkins, quipped, "I call them their parkas when there are lots of them."

Marie Alexie (June 2016:121) also mentioned that people used cottongrass as an indicator that *cingikegglit* (humpback whitefish) were starting to come out of the lakes into the main river: "Then during the berry-picking season, when cottongrass tops started to turn brown, the fish started to come out."

Melquruat (white cottongrass).
KEVIN JERNIGAN

Annie Jackson (July 2000:29) noted that although cottongrass may not appear to be edible, it is: "If you want to really cleanse your stomach, just start off making *akutaq* and put these in, that white stuff before it starts flying. And this will really clean your insides. And here we think they're a nuisance, too. When we go out berry picking, all we have to do is clean this [stem] from the salmonberries, but these [cotton pieces] are edible."

George Pleasant (January 2011:280) added a short story: "There is lots of [cottongrass] in our village. One day, when the people building a road took a coffee break, when their boss entered, Joe Carter evidently told him, 'There's lots of cotton out there. You can make a sweater out of it.'" Everyone chuckled.

Ciilqaaret
Fireweed

Ciilqaaret (fireweed plants, *Epilobium angustifolium*) are common throughout southwest Alaska. Like wild chamomile, they favor disturbed areas. In spring, people gather the young, edible shoots, and later in the summer pick the flowers, which are eaten raw, and leaves, which can be made into tea. Annie Andrew (July 2000:30) remembered:

Before these ripen and when they are still small, they are eaten. They say not to eat too much when they are small. They say it can upset a stomach when they are eaten on an empty stomach, but if you eat them with something, it will be okay. Those who died before us also ate these

Ciilqaarat (fireweed). KEVIN JERNIGAN

Young, edible fireweed shoots.
ANN FIENUP-RIORDAN

newly grown ones with seal oil. These are like medicine, and they can clean your body and [take away] your sickness.

Annie added that, like wormwood, fireweed can be used as a poultice to heal a cut or a boil.

Anna Agnus (July 2007:622) described the healing properties of fireweed tea: "They call the ones with purple flowers *ciilqaaret*. In the past, our parents would use those as tea. They would gather them when they wilted. And I would also gather them. . . . It is said these work on a person's blood and bones. I believe in their ability to heal. I'd drink it in the morning and at night after filling a cup." Annie Andrew (July 2000:30) cau-

A field of tall fireweed near Scammon Bay. JACQUELINE CLEVELAND

tioned: "They say these are strong, and they say not to drink the liquid from this all the time. . . . There is a saying that a lot of this shouldn't be taken."

Along with their ability to heal, women also described playing with fireweed when they were young. Laughing, Cecelia Andrews (April 2019:264) remembered: "When we were playing outside, we would pick them and do things. We'd become flowers, we would Yup'ik dance." Ruth Jimmie added:

> We would braid them before they grew flowers and pretend they were our skirts, pretending to be Hawaiians and dance.
> And the boys would use them as spears. Our past was so joyful. There was nothing [modern] around.

Ikiituut tarnat-llu
Wild celery and cow parsnips

Ikiituut (wild celery, *Angelica lucida*) is another well-known plant in south-west Alaska. Many recalled peeling and eating the hollow stems of young plants while out on the land. Pauline Matthew (January 2011:249) noted: "After peeling them like celery, they put them in seal oil and ate them." Martha Mark added that wild celery were eaten in June, when they were newly grown. Theresa Moses (May 2004:71) noted: "When the *ikiituut* first started to grow, they gathered them. And when [the stems] got solid, they could no longer [be eaten]." Eating the stems of young plants was considered generally healthful.

The dried roots could also be crushed and used to treat body aches (Jernigan et al. 2015:101). Theresa Abraham (February 2019:30–31) described how once, while picking berries, her cousin had instructed her to rub the fresh root of a wild celery plant on her back for pain relief: "Just like medicine, my back started to get hot, and it felt a little numb; and that area on my back got better while I was picking berries." Theresa's cousin also warned her not to use wild celery roots in the steam bath, as they are too strong and might burn. People distinguish between male plants with flowering tops and female plants without tops. Cecelia Andrews (April 2019:241) said that people only used the roots of male plants as medicine.

Simeon Agnus (July 2007:619) described burning wild celery roots as mosquito repellent: "Their roots are also great to use as mosquito repellent, after drying them first. And their smell is much more pleasant than [store-bought] repellents. . . . They didn't have a bad odor as they slowly burned down, constantly giving off smoke. If we don't have any store-bought mosquito repellent, the roots of wild celery plants are readily available. That's how they are used."

Like white cottongrass, wild celery was used as an indicator that *cingikegglit* (humpback whitefish) were leaving lakes in the fall. Nick Pavilla (December 2014:115) of Atmautluak noted: "When the top of the wild celery began to bloom and snapped open [at the end of June] they'd say, 'The *cingikegglit* are turning back now.' They'd be the first ones to swim downriver to go out [to the Kuskokwim River]."

When discussing wild celery, elders also mentioned *tarnat* (cow parsnips, *Heracleum lanatum*). Cow parsnips are similar in appearance to wild celery but grow in different areas. Martha Mark (January 2011:250) noted: "Cow parsnip grows in the mountains in our home area [Quinhagak] unlike wild

➤ *Ikiituk* (wild celery) in flower, too late to pick for food.. KEVIN JERNIGAN

Fannie Cleveland Moore harvesting wild celery, June 5, 2019. JACQUELINE CLEVELAND

Ikiituk (wild celery). KEVIN JERNIGAN

celery that grows [around the village]. But in Platinum and Goodnews Bay, they get cow parsnip from behind their village since they are next to mountains." Like wild celery, people peel and eat the stems of young cow parsnips before they bloom. Denis Shelden (January 2013:339) noted: "Their stems when they first grow, before they get hard, are good eating after peeling them." Raphael Jimmy replied: "They are very delicious with seal oil. At our camp in spring, there are a great many of those *tarnat*." Jernigan (et al. 2015:69) notes that the leaves should not be eaten, as a chemical in the surface hairs can numb the lips, and some people have a strong allergic reaction to it. John Andrew (April 2019:244) remembered seeing people using large cow parsnip leaves to cover their barrels full of berries, but these leaves were not eaten. John added: "Bears feed on cow parsnips, mostly the male [plants], before the salmon arrive. Maybe they call them cow parsnips because cows feed on them. Bears feed on them, too."

Like wild rhubarb, the dried stalks of both wild celery and cow parsnips were said to have stood as high as people in the past, when frost covered them in cold, windless weather.

Tarnat (cow parsnips). KEVIN JERNIGAN

Tarnat (cow parsnips). KEVIN JERNIGAN

Poison water hemlock (*Cicuta mackenzieana*) is another plant that resembles wild celery. Bob Aloysius (October 2010:131) commented: "That [wild celery] that was just growing, we harvest those, before they get hard, when they're nice and tender. But I am afraid of them at this time since those darn things, those *uquutvaguat* [poison water hemlock] look similar, and they're poison." As noted below (page 162), water hemlock can be distinguished from wild celery by its smaller leaves as well as its spherical top, compared to the flatter tops of wild celery. Also, the roots of water hemlock have larger chambers than wild celery roots and are very poisonous.

In the past, the smoke of wild celery, like the smoke of *ayuq*, was used to purify a person's body as well as sod homes and the *qasgiq* (communal men's house). Cecelia Andrews (April 2019:35) said that wild celery smoke could make a ghost disappear. Brentina Chanar (June 1989) of Toksook Bay noted that a nursing mother with a newborn boy was prohibited from eating wild celery, because the plant was considered to have "female flesh." Theresa Moses (August 1987:7) described how a shaman gave her wild celery as a *napan* (lit., "something that keeps one upright") when she was sick as a child.

Cow parsnip. RICHARD W. TYLER

Subsequently, she could not eat wild celery, but grew fat and thin along with the plant each spring and fall.

In the past, wild celery played an important role in Yup'ik ceremonies, especially the Bladder Festival. The Bladder Festival was held each year to honor the *yuit* (persons) of the seals and other animals, said to reside in their bladders. During the festival, bladders of seals taken during the previous season were inflated, hosted for a number of days, then pushed through an ice hole and sent back to their homes in the hopes that they would return the following season.

The Bladder Festival proper began when two or more young men were sent out to gather wild celery, which they treated as sentient in a number of ways—leaving gifts of food and water at the plants' base before gathering them, as well as hosting them in the *qasgiq* after their return, placing them over the doorway in the seat of honor. Francis Charlie (January 2013:343) described men gathering wild celery, along with their roots: "When the people were going to have the Bladder Festival, they [gathered it]. If a person took as much as he could, he would bring it home. For some reason, they say when some people would pull [an *ikiituk*], it would come out with its roots although the ground was frozen." Wendell Oswalt (1957:31) reported that along the Kuskokwim, people believed that the roots of wild celery "represent each man's partner from the underworld and appear to represent spirits of the dead." If so, the ceremonial sequence of the Bladder Festival may have enacted a relationship between ancestral spirits and the *yuit* of the seals. Just as the *yuit* of the seals were drawn to and hosted by human hunters, so were the human ancestral spirits—in the form of wild celery roots—who also received gifts later in the ceremony (Fienup-Riordan 1994:279).

During the Bladder Festival proper, men regularly purified both the bladders and the human participants with the aromatic smoke of wild celery, which was said to please them (Nelson 1899:393). Then, after the wild celery had dried, men erected a ceremonial pole, the *kangaciqaq*, to which stalks of wild celery were attached. Throughout the Bladder Festival, the *kangaciqaq* was feasted and hosted like a living person. At the close of the Bladder Festival, the *kangaciqaq* was lit and lifted through the skylight, along with the bladders that had been hosted during the festival. Carrying both the bladders and burning *kangaciqaq*, men ran to a hole in the ice and pushed the bladders in so that they could travel to their underwater home, where the people hoped they would boast of their good treatment and return the following season (Fienup-Riordan 1994:292–96).

Anuqetuliar
Yarrow

Elders we worked with commented only briefly on *anuqetuliar* (yarrow, from *anuqa*, "wind," *Achillea millefolium*), also referred to as *anguqetussngit* on the Kuskokwim. John Andrew (April 2019:234) quipped: "Sounds like *anuqessuucet* [wind banners] because they're always moving when there's a breeze." Elizabeth Andrew (April 2008:206) noted that one should chew the roots and then swallow one's saliva to treat mouth sores or sore throat. When chewing, she said that the mouth would feel numb—an indication that she is talking about yarrow, not Siberian yarrow (*Achillea sibirica*), which Jernigan (et al. 2015:99) said does not have this numbing characteristic. Elsie Tommy (March 2009:311) noted:

> Long ago when people no longer had tobacco, they would gather the roots of those small plants and put them in their mouths. After chewing them, they would start to have a tingling sensation in their mouths; they used to chew them, likening [the sensation] to tobacco. . . .
>
> They didn't use their tops, but only their roots.

Martina John (March 2009:312) added that their roots are small and short: "They cause numbness like when chewing tobacco." Jernigan (et al. 2015:77) noted that in some areas, people chewed yarrow roots to treat sinus infections, asthma,

Yarrow. SHARON BIRZER

Anuqetuliar (yarrow). KEVIN JERNIGAN

and pneumonia. Others made tea from the younger plants. Some boiled the whole plant and used it as a hot compress to treat arthritis, or used the steam to treat coughs and colds.

Qaltaruat wall' pellukutat
Coltsfoot

Coltsfoot (*Petasites frigidus*) is known as both *qaltaruat* (lit., "pretend *qaltat* [buckets]") and *pellukutat*. Emma White (April 2008:217) of Quinhagak described the use of dried coltsfoot leaves to treat diarrhea and stomach pains:

> When I had severe diarrhea and stomach ache, someone added that plant to a cup of boiling water and had me drink the entire thing. I immediately recovered. That medicine is wonderful. . . .
>
> The same person always gathers them. And at Quinhagak, when we have the stomach flu, we go to her. You collect them in the summer and dry them, and when they're dried, you store them.

Qaltaruat (coltsfoot), also known as *pellukutat*. KEVIN JERNIGAN

Coltsfoot. SHARON BIRZER

Qaltaruat (coltsfoot). KEVIN JERNIGAN

> When she stores them away, they turn brown, and then she takes
> some from there, fills a cup and lets it boil in the microwave, and then
> after cooling it, she has us drink it. It is really good.

Along the coast, coltsfoot leaves were used to make *peluq* (ash) to mix
with chewing tobacco to enhance its strength. Theresa Abraham (March
2008:227) recalled: "When those *qaltaruat* wilted, some people gathered
them and made ash to mix with tobacco, even in spring." Francis Charlie
(January 2013:348) remembered being sent to gather the leaves: "When they
wilted, my grandmother used to have us get some to burn. They are no different from *kumakaq* [punk]. They are strong." Theresa Abraham (February
2019:118) added that their ashes are white like the ashes of *kumakaq*.

Francis Charlie (January 2013:348) also recalled the use of these large
leaves as berry buckets for children, hence their name *qaltaruat* ("pretend

buckets"): "They would grow during the time salmonberries grew. And when we were small, we would sew them [closed] with a blade of grass and use them as containers when we picked salmonberries. . . . A blade of grass could make a hole in it. When you made a hole, then you [sewed it], then you let go. It could then be a container [without a handle]." Ruth Jimmie (February 2019:118) agreed: "We would just fix them and use them as buckets when we picked berries."

Teptukuyiit
Valerian

Teptukuyak (valerian, from *teptu-*, "to be odoriferous," *Valeriana capitata*) was also recognized as having medicinal value. It could be used either fresh or dried and stored for later use. Theresa Moses (September 2003:224) had seen her mother putting valerian leaves in her nostrils in winter:

> I asked her like this, "Why do you have your nose plugged with that?" She said that when she kept getting headaches she plugged them. . . . She told me that they were small pieces of *teptukuyak*.
> When she kept getting headaches, she would stuff it in her nose and her headaches would disappear. She would smell the scent of the plant all the time.

Fannie Jacob (March 2017:283) of Napaskiak noted that her grandmother also used valerian leaves to relieve headaches, wearing them in her hair. Jernigan (et al. 2015:114) notes that others boiled the roots to make a calming tea to relieve anxiety or breathe the scent to reduce throat congestion. John Andrew (April 2019:247) said that his wife added their leaves to salmon soup for flavoring: "Their small leaves are good added to cooked food."

Alexie Nicholai (March 2016:145) described using *teptukuyiit* to make his nets better at catching fish: "Those before us took things from the ground and tied them to the lead line of their fishnets. They make them better at catching; they have some sort of scent." Jacob Black (March 2016:145) had used wormwood in the same way. John Andrew (November 2018:33) explained:

> After driftnet fishing if they don't catch anything, they would look for *teptukuyiit* with small purple flowers at the top. They would dig them

➤ *Teptukuyak* (valerian). KEVIN JERNIGAN

out; underneath, their roots smell very strong. They would [tie valerian roots] at the middle and at the ends of the net.

They said that when they would fish with their drift net, the fish would probably smell and be attracted to its odor.

They are pretty little purple flowers. They aren't large; they have a lot of roots underneath, and some would have a small bulb. They used that kind of [plant] to *iinruarturluki* [apply medicine to their nets]. . . .

Their roots look like fake hair when you pull them, there are lots of them. People who had a hard time catching would tie them on the middle of the gill net and on the end.

Caqliit
Roseroot

Caqliit (roseroot, *Rhodiola integrifolia*) are also known as *megtat neqait* or *evegtat neqait*, both translating as "bumblebee food," as these small insects favor their flowers. Cecelia Andrews (April 2019:39) recognized them as growing near her home: "Those *caqliit* [roseroots] grow on the marsh. Their tops are red, and their roots are very white when opening them." Nick Andrew (February 2019:93) described their roots as tasting like turnips when eaten raw in spring: "I used to like them, I would pick them up, throw [the top] away, and eat the roots." Ruth Jimmie (April 2019:74) remembered eating only their flowers, which tasted sugary.

Both Cecelia Andrews and John Andrew (April 2019:39) described using them to treat cuts and canker sores in the mouth, and as a treatment for ulcers. Cecelia remarked:

These are medicine. They take the roots and spill hot water on them. Any cut or sore will heal right away. . . .

I tried picking those. And after cleaning them, I put them inside a glass and spilled boiling water over them. That turned a really nice yellow color. One has to take one small spoonful. I know it will clean my insides for sure.

Caqliit (roseroot). KEVIN JERNIGAN

Uqviit, enrilnguat, cuyangaaraat-llu
Willows, young willow shoots, and willow leaf buds

The inner bark, leaves, and leaf buds of a variety of willows (*Salix* spp.) were widely recognized as possessing medicinal qualities. Anthropologist Margaret Lantis (1959:5) noted that both willow leaves and willow bark contain the principal component of aspirin—salicylic acid—giving people some analgesic relief. Willows generally go by a variety of names, many of which are variations on *uqviit*, including *uqvigaat* and *uqviaret*, as well as *uqvigpiit* (Alaska bog willows, lit., "real *uqviit* [willows]," *Salix fuscescens*).

Annie Andrew (July 2000:26) said that *enrilnguat* (young, edible willow shoots, from *eneq*, "bone," lit., "ones similar to ones without bones" because they are bendable, *Salix* spp.) are good for mouth sores: "They are cooked into a strong brew. If a child or adult has sores in the mouth, use a very white cloth to scrape the sores with the liquid, even though it hurts. You must remove all the sores and let it bleed. Then after it is boiled into a strong brew, put the cloth on top of the sores, even though it stings, and the sores will heal." Joshua Phillip (December 1985:141) agreed: "They used it for sores inside the mouth. If you had mouth sores, you were told to eat the soft willow shoot, removing its bark. If he chews on the inner skin of *enrilnguat*, his mouth will get numb immediately."

John Andrew (February 2019:202) briefly recounted healing mouth sores with *uqvigaat* that was not young:

> One time in Anchorage, I had sores in my mouth. I noticed there was a willow outside that University Center. I went over and broke [a branch] off. [Someone said to me,] "What are you doing, you're damaging it." [I replied,] "No, I'm going to use it for medicine." He just stared at me.
>
> Then in front of him, I just chewed on it and kept it in my mouth. The next day, my sores were gone.

Enrilnguat (young willow shoots).
TORRE JORGENSON

Annie Jackson (July 2000:26) noted that *enrilnguat* can be used to treat abdominal problems: "If you have abdominal problems like swelling or kidney trouble, you can eat this raw, the inner part and the outer part together. Just chew it and eat and swallow it. That's the medication that can cure the swelling." Annie noted that *enrilnguat* bark can also be cooked and used to treat pneumonia and bronchitis: "Peel off [the bark] and take the inner part and make strips out of it and cook it and boil it for at least five minutes. Then cool it. And you can take this medication. It's sour tasting. But it can cure your lungs, your abdomen, your kidneys. It's really good for those as a medication."

People also ate the leaves and leaf buds of *enrilnguat*. Wassillie B. Evan (December 1985:141) recalled: "If you have very bad heartburn or indigestion, add willow leaves to your cooking. If someone eats them and swallows the juice, the heartburn will go away. When I have heartburn, I go to the willow trees and chew on the leaves." Joshua Phillip (December 1985:103) noted that their buds are good to eat with processed salmon eggs. Peter Gilila (February 2019:114) described picking the young leaves and making them into *cuassaat* (cooked wild greens). Nick Andrew (April 2019:7) also mentioned picking the young leaves of *cuyaqsuut* (diamond leaf willows, lit., "ones with good leaves," *Salix*

Dwarf willow with catkins. TORRE JORGENSON

pulchra) and eating them with seal oil. John Andrew (April 2019:184, 194) was enthusiastic about this abundant spring vegetable: "Far upriver, we eat *enrilnguat*; they are delicious. Before their leaves harden, they add them to cooked food and even to half-dried salmon and cook them. Or they cook them on their own and soak them in seal oil. . . . And when they used to make aged salmon roe, women would gather them and eat those along with aged salmon roe when they didn't cook them."

Mark John (April 2019:36) described eating *uqviggluk* (pussy willows) in spring: "The things that grow on the tips of *uqvigaat* [willows] with a lot of *melqut* [fur]. . . . We would take those and eat them at Umkumiut when we'd play." Ruth Jimmie also remembered chewing on willow tips, pretending they were gum. Peter Gilila (February 2019:50) recalled seeing some people remove their fur, while others ate them as they were.

Martha Mark (January 2011:274) described using *cuyangaaraat* (willow leaf buds, lit., "early leaves," *Salix* spp.): "When I cooked bird soup, when we were up [in the mountains], I picked some *cuyangaaraat* from outside. And after washing them, I added it to soup. It was delicious."

Kaviqsuut (little tree willows). KEVIN JERNIGAN

Olinka George (April 2008:204) of Akiachak described boiling the bark of *kaviqsuut* (probably red-barked or little tree willows, from *kavir-*, "to be red," *Salix arbusculoides*) and drinking the juice periodically to treat colds: "We use their bark, and we also cut up their branches. I boil them for ten to fifteen minutes. When we boil them for fifteen minutes, they get dark and their water is reduced, and then we add [more water] to it. [Then I drink] just a mouthful, two [or] three times a day."

Matthew Bean (December 1985:103) of Bethel noted that sprouts of *cuyakegglit* (diamond leaf willows) were picked in summer and stored in seal oil: "Then in the fall, they would eat them with fish." Raphael Jimmy (January 2013:335) described their medicinal uses: "For those of us on the coast, those *cuyakegglit* [are medicine]. After removing their bark, their small membranes on the inside [inner bark], when one has a sore throat, or when one has sores in his mouth, they say after [removing those] and chew-

Cuyakegglit (diamond leaf willows). KEVIN JERNIGAN

ing them, it heals one who has a sore throat or one who has grown mouth sores. They cause [the sores] to sting."

Francis Charlie (January 2013:364) described using the juice of *cuyakegglit* bark to treat snow blindness:

> When I was small, my grandfather never was without the bark of *cuyakegglit*. Sometimes when he was ill, he would fill a spoon from a bottle and drink a little. And he never gave me any.
>
> But I tended to get snow blindness back in those days in spring. [My grandfather] would put drops of the juice of *cuyakegglit* in my eyes. It was very painful. The next day I would find that my eyes had already stopped stinging.

Later, when Francis began to cough and spit blood, his grandfather used *cuyakegglit* juice to treat him:

> Starting from there, in the morning and at night when I was about to go to sleep, although I didn't like the taste, he would have me swallow [the juice]. Since it was stronger than tobacco [juice], although I thought to spit it out, since he seemed to be watching me, I would swallow it.

Although I didn't spit blood like that again, he didn't let me quit [taking the juice] right away, but had me take it for about a month.

Years later, Francis had an X-ray that showed that, although he had not been aware of it, he had tuberculosis in the past: "I was thinking that the *cuyakegglit* juice had healed my TB."

Elngut
Birch trees

Elngut (birch trees, *Betula papyrifera)* have many uses. Their wood is fuel, their bark provides material for baskets and canoes, and straight-grained pieces can be split and made into bows. John Andrew (April 2019:214) recalled: "The ones that aren't dried don't break easily. That's why they call them *elngut* (lit., "pliant ones," from *elngur-*, "to be thick, but pliable").

Annie Andrew (July 2000:26) described using the inner bark of *elngut* in a way similar to willow bark as medicine to treat pneumonia and tuberculosis: "What we call inside meat [inner bark], it can be a medication, especially if you get TB or cancer or pneumonia. You can cook this and drink it every day, not in large amounts, but one-fourth of a cup will do. . . . It's really sour when it's cooked. This inner part . . . is good for your health." Peter Gilila (February 2019:25) said that the inner bark or "meat" of birch trees was also considered edible in an emergency:

> I remember that one time Mic'imalria and I, our outboard motor broke down when we were heading home after going to spring camp. Since we didn't have provisions in the spring, he split open a birch tree, removed the bark, and he let me put a little bit in my mouth. He said my body might start to get too weak, that I should put some of that in my mouth. Even though I didn't know what it was, it was evidently *elngum kemga* [the birch tree's meat].

John Andrew (April 2019:213) described the healing properties of the tips of young birch branches:

> One elderly woman told me that the tips of these young [birch] are medicine. She said she would look for the young [birch] and cut off their tips to make medicine by cooking them. Then she would take [the liquid]

Elnguq (birch tree). KEVIN JERNIGAN

as medicine, drinking a little bit. She said they make you feel better; they make you more lively. In spring, she said their newly grown parts, their tips are semi-sweet.

Birch sap was also considered medicine. Peter Gilila (February 2019:33) recalled:

There are many birch trees in spring; my grandmother would have me tear the bark and insert a small nail on it. Making a small opening on the birch tree, I would poke the nail in, and she'd have me place a small container down below it. It would drip slowly. And when [the container] had a certain amount inside it, I would bring it over. I would eat it, adding it to fry bread. She said it's not just used for that; she said it is medicine, too.

△ *Phellinus igniarius* harvested by James Nicholai and burned to make punk ash. JAMES NICHOLAI

◁ Birch bracket fungus. JAMES NICHOLAI

Finally, Peter Gilila (February 2019:119) noted that *kumakaq* (tree fungus, including birch-bracket fungus and *Phellinus igniarius*) can be boiled and the resulting liquid used as medicine: "They used to boil the *kumakat* of birch trees and have them as a warm drink. My grandmother used to drink them as a hot beverage once in a while. After boiling it, they drink its juice. She used to have me drink the water from *kumakat* once in a while, after boiling it."

Cuukvaguat wall' auguqsulit
Alders

Cuukvaguat (alders, lit., "pretend *cuukvak* [pike]," *Alnus viridis* subsp. *sinuate* and *Alnus incana* subsp. *tenuifolia*), also known as *auguqsulit* (possibly from *auk*, "blood"), were recognized as having medicinal qualities. Peter Black (April 2012:64) originally from Hooper Bay, described using alder bark as medicine. Joe Felix (July 2007:629) of Toksook Bay added detail: "Alder bark is good for those with sore throats. They also give them alder bark after boiling it. When our grandchildren have sore throats, that's what we have them do. We just put a little bit in their mouths. I think it numbs the throbbing pain [in the throat]. We also cook alder bark and use it like that."

Cuukvaguat
(alders). KEVIN
JERNIGAN

Elizabeth Andrew (April 2008:219) described eating the light-colored growths on alder bark to treat diarrhea: "On old *cuukvaguat*, you'll see something stuck to [the trunk], I think it's yellowish in color. If one picks those and washes them, they say when a person having diarrhea eats them, they'll stop having diarrhea."

Kevraartut wall' nekevraartut
Spruce trees

Kevraartut (white spruce, *Picea glauca*) also known as *nekevraartut*, have a number of medicinal uses, especially their needles. Many people use spruce needles in the steam bath. Elizabeth Andrew (April 2008:205) said: "When taking steam baths, we [put them on the hot rocks and] let them give off smoke along with steam, but we don't let it smoke too much." James Guy (June 1995:21) added: "Shake off the [needles], dry them, and then put them in a bag. They are good used like incense, to help breathing." Joshua Cleveland (January 2011:288) noted that people on the treeless coast are also beginning to make use of spruce needles in this way: "Now that we [here in Quinhagak] and the people of Eek have started to get spruce trees for firewood, they have learned to strip [the needles] from the branches, and when they take steam baths, they place them on the rocks and have them smoke."

Peter Gilila (February 2019:37–39) spoke of rubbing softened spruce needles directly on one's body: "After they rubbed [the needles] in their hands in a circular motion really hard, they would put them on their bodies, probably when they had problems with their breathing. After doing that, they would feel better. This is when they have lung issues."

Spruce needles can also be boiled and the resulting liquid taken as medicine. Annie Andrew (July 2000:28) explained: "If this is made for a medicinal drink, it will be strong. They always say not to drink too much. This is medicine for us Yup'ik people." Joshua Phillip (December 1985:141) declared: "I cooked the spruce needles and drank the juice, and it cured my TB." Elizabeth Andrew (April 2008:205) mentioned boiling spruce needles and wormwood together, and drinking the juice to treat cancer and other ailments.

John Andrew (April 2019:218) noted that his mother cooked spruce branches and drank the juice for general good health:

> They used their small branches as medicine, as a juice to drink. I used to add their small cones to them; the ones you added cones to were very strong. . . .
>
> When my parents were alive, my mother used to tell me that I should go and get some spruce branches to cook, for her to prepare medicine. She also drank it and took it as medicine like they do these *caiggluut* [wormwood].
>
> Just a small amount, sometimes in the morning and at night. They said taking too much can cause a stomachache.

Kevraartuq (spruce tree). KEVIN JERNIGAN

Peter Gilila (February 2019:37) agreed that people were admonished not to take more than a teaspoonful at a time to alleviate cold symptoms.

Annie Andrew (July 2000:28) noted the medicinal use of spruce pitch: "Spruce pitch can also be used as medicine for anything. The sticky tree pitch can be used on infections and cuts. The sayings of these [spruce] are many." John Andrew (April 2019:216) added that people also gathered and ate spruce tips before they hardened, both raw and cooked. John (April 2019:217–20) noted the many additional uses of spruce: spruce were gathered as firewood and their roots used as lashing, plus spruce boughs provided excellent insulation when camping in the wilderness: "When the ground was frozen, we also put the branches of this tree as insulation, and after putting a tarp over it, we would place our things on top. Ones prepared in that way were soft, using them as mattresses. And our clothing and things smelled nice."

Though spruce trees were largely beneficial, they were considered dangerous to all *caagnitellriit* (those experiencing life transitions, including first menses, death in the family, or miscarriage), and people in that situation were warned to stay away from them. Nastasia Larson (March 2017:191) added a caveat: "If the person accidentally found himself next to a spruce tree, he was told to take a needle and place it inside the garment and say, 'I am now part of you.' . . . I've heard that spruce needles are powerful because they are sharp." Thus, these dangerous needles could also be used as a weapon of protection, to keep harm away. Elizabeth Stevens (April 2015:307) of Napaskiak recalled: "When a person who is following *eyagyarat* [abstinence practices surrounding life transitions] comes upon a spruce tree, he takes [a branch]. Then he puts it inside his pocket or [underneath his clothing, touching the skin]. . . . They say if we don't put it inside, its needles will poke our bodies. Later, it will become an ailment or source of pain."

Qamlleq wall' araq
Ash

Burned wood ash, known as both *araq* and *qamlleq* (from *qame-*, "to die down [of fire]"), was also said to have medicinal properties. Joshua Phillip (December 1985:141) gave a detailed explanation of the process of preparing ash as medicine:

When the joints get big from arthritis, the ash from burned wood is a medicine. The ash from the stove. Put some in a cooking pot, cover the pot, and boil it, then put it aside and cool it. Then when the ash settles, strain off the ash and cook the juice again. When you do that, a white substance settles to the bottom of the pot. Then pour off the juice again and pry off the white stuff, which looks like rock but not too hard, stuck to the bottom of the pot.

Then boil the juice a third time. Then finally, it will be very clear with nothing at the bottom. Then when it cools, you can dip the joints that have pain. After soaking your joints several times, they will break out and look like they are getting sores. But continue to soak them.

They say that will pull it [out]. The joints that had gotten big on his hands would become normal again. They say that is very strong medicine.

Araq (ash) made by James Nicholai from birch bracket fungus. JAMES NICHOLAI

People also burn *kumakaq* (tree fungus or "punk," from *kuma-*, "to be lit or burning") also known as *arakaq* (lit., "ones that will be turned into *araq* [ash]") on the Yukon, to produce ash that can be used in the same way as wood ash to treat sore muscles or steam the body. *Peluq* (ash from either *kumakaq* or willow) can also be mixed with *cuyat* (tobacco leaves, *Nicotiana tobacum*) to strengthen their effect when used as chewing tobacco. *Iqmik* (tobacco leaves mixed with ash) is more potent than tobacco on its own, as ash is alkali and helps convert the nicotine in tobacco into its freebase form, which the body can more readily absorb.

People living in areas with willows sometimes burn large quantities and sell them to people living in areas where willows are less common. Paul John (October 2002:223) recalled: "I would gather *uqvigpiit* [Alaska bog willows]. [In years past,] I would make some for those around us. I had a big place to burn them. I would sell them at Chefornak, Nightmute, and Tununak, and they would sell all at once. I would earn over a thousand dollars." Paul noted that when adding willow ash to his chewing tobacco, it increased its potency. Some, like Paul, prefer willow ash, while others prefer ash made from *kumakaq*.

Denis Shelden and Francis Charlie (January 2013:350) also discussed the use of birch or willow punks—known both as *arakat* and *kenerkat* (lit., "kindling, fire starters"). Francis stated that *kenerkat* burn slowly and stay lit. Denis added: "They say they used to cover them with ash and bring them to places. And when they wanted to cook in the fire, they would take that out and blow on it, and it would suddenly light up." Francis noted that if the burning punk was covered in ash, it would stay lit for twenty-four hours. Denis recalled another use: "They're like Buhach, like coils [burnt to keep insects away]. Mosquitoes don't like to be near their smoke."

Peter Gilila (February 2019:73) described the use of *kangipluk* (charcoal) as a hot compress to treat sore muscles:

> While *kangipluk* was warm, he would put it in a container when I had throbbing pain; when I hurt my back, he used to have me keep it pressed on there. He would heat the charcoal from the stove, and not have it be completely hot but very warm, and put it on there. After doing that for a number of days, my pain went away. But he would put it in cloth. . . . He didn't add water to it, but left it as it was. . . . Even *kumakat* [punk] . . . or a rock [could be used in that way].

John Andrew (April 2019:259) noted that warm wood ash could be used in the same way.

Finally, along the Kuskokwim, when a young girl menstruated for the first time, she was instructed to wear a small pouch containing, among other things, a piece of charcoal. Those who were abstaining following a death in the family might also carry ash or charcoal as protection when traveling. Ralph Nelson (April 2015:308) of Napakiak noted: "I also heard about charcoal from a fire; they made something [like a pouch] and had the man who would travel wear it." Alexie Nicholai (April 2015:317) of Oscarville explained: "It's because these *ircenrraat* [other-than-human persons] really like to mess with people. When a person who is abstaining travels, he would [leave] after getting ash. They say those [*ircenrraat*] cannot see them, or they are foggy [not clearly visible]." *Ircenrraat* are known to have poor eyesight, and Ralph noted that both ash and charcoal make it difficult for them to see. Alexie (March 2017:268) later said that ash might also be sprinkled on oneself or one's hunting equipment with the same effect. Jacob Black (August 2017:58) agreed that ash blinds *ircenrraat*, and that those in that situation can scatter ash on boats and on the surface of the water as protection.

Agyat anait (puffball mushrooms, lit. "star feces"). KEVIN JERNIGAN

Agyat anait
Star feces

Several useful species of mushrooms grow in southwest Alaska. Probably the most noteworthy are puffball mushrooms (puffball fungus, *Lycoperdon* spp.), known as *agyat anait* (lit., "star feces"). Theresa Moses (September 2003:232) described their origin as falling stars: "When we see them in space, they are white as they travel [as falling stars]. They say that these [puffball mushrooms] are star feces. Back then they said that there were many of these, maybe when [the stars] used to *anaq* [drop feces]. When we watch them, they are lit as they travel [through the sky]." Wassillie B. Evan added: "They are seen when they want to be seen [on the ground], and when they are dried and touched, they smoke from their contents." Wassillie also noted that puffball mushrooms are found on the land in different sizes. Elizabeth Andrew (April 2008:215) recalled how the brown powder inside *agyat anait* can be used to treat sore throats or canker sores: "If you sprinkle the contents of a puffball mushroom [in your mouth], it will heal right away." In fact, Jernigan (et al. 2015:154) notes that the species *Lycoperdon perlatum* has been investigated for its antimicrobial properties.

TUQUNARQELLRIIT
NAUCETAAT
POISONOUS PLANTS

✳

Although few plants are poisonous in southwest Alaska, children are warned from an early age to recognize them and to avoid them. Incidents where people have mistakenly eaten them and suffered the consequences continue to be passed down. Plants eaten by swans and geese, however, are not deadly. John Andrew (September 2018:17, April 2019:26) shared what he learned when young:

> They told us to observe what swans eat in the wilderness. They said that if you don't have provisions and a person eats what they eat, he will not die. They also say that a person can eat what various geese eat. They say that there are no poisonous plants [in their diet]. . . .
>
> And they say a person can also eat the food that mice picked and gathered. They told us to watch the foods of these animals from the wilderness, that if we don't have anything [to eat] we can [eat them]. And they'd gather large amounts not just for themselves; they'd have them distribute them to people with no providers when they arrived home.

Uquutvaguat wall' anguteurluut
Poison water hemlock

Chief among poisonous plants is water hemlock (*Cicuta mackenzieana*), known as both *uquutvaguat* and *anguteurluut* (lit., "poor, dear males"). While water hemlock resembles wild celery, it has smaller leaves and is found growing with its base submerged in water. Also the tops of water hemlock are more spherical, while wild celery tops are flatter. The roots of water hemlock—which have larger chambers than wild celery roots—are very poisonous, although muskrats and beavers consume them with no ill effects. Albertina Dull (November 2017:2) declared: "That type of food is fatal. But the mice or muskrats that eat them don't die. . . . [Water hemlock] grows along the shores of lakes and looks like wild celery." John Andrew (April 2019:31) added confirmation: "The poison hemlock tops almost look like *ikiituut* [wild celery]. But if you pull their roots out, those poisonous ones have bulbs, and some of their bulbs are large. They tell people not to eat those because they are deadly. . . . They said if a person eats them, he won't be alive."

Lizzie Chimiugak (June 2009:155–58) shared a short *quliraq* (traditional tale) in which a married couple came upon a large den on top of a mountain. The man took a light into the den and traveled along. First, he met a tiny shrew, and he asked it, "Where is the scary one?" The shrew answered, "In there, at the source." The man continued to travel for a long time, meeting larger and larger fur animals and asking them all the same question. Finally, he arrived at a door where he found a person lying on his back along the wall with his stomach slashed open. The man dropped his light in fear and quickly ran out of the den. When he emerged, he found that his house had rotted and his wife was gone. Bending over a small lake to drink, he saw that his reflection was that of an old man. Lizzie concluded:

> Since he didn't know what to do, he suddenly fell on his butt along the shores of a lake in the water, and he became an *anguteurluq* [lit., "poor, dear man"]. Those *anguteurluut* that grow that have poisonous tubers; as you know, lakes have them.
>
> He suddenly sat down there. "Let me be one of these forever, so that in the far future, they will harvest me."

➤ A stalk of *uquutvaguaq* (poison water hemlock), also called *anguteurluq*. Water hemlock can be distinguished from wild celery by its smaller leaves as well as its spherical top, compared to the flatter tops of wild celery. The roots of water hemlock have larger chambers than wild celery roots and are very poisonous. KEVIN JERNIGAN

Pupignat
Poisonous mushrooms

Pupignat (poisonous mushrooms, lit., "ones that cause sores") also grow in southwest Alaska. Nick Andrew (February 2019:20) said: "Their tops are very red. They tell people not to eat those; they say they are poisonous." John Andrew (April 2019:30) described observing men eating small quantities of red-topped mushrooms (probably *Amanita muscaria*) for their hallucinogenic properties when traveling in the mountains: "And in the trees there are ones that are red with white spots. They said those can cause one to become intoxicated. We try not to tell young people about those, afraid that they may try to use them to get drunk."

A *pupignaq* (poisonous mushroom), *Amanita muscaria*. ALASKA BOTANICAL GARDEN

Amanita muscaria. SHARON BIRZER

Tulukaruut neqait
Baneberries

....

Tulukaruut neqait (baneberries, lit., "raven's food," *Actaea rubra*) are also poisonous. Elizabeth Andrew (April 2008:215) noted: "Those berries in trees in a bunch that are red; they say those are food that ravens eat and they say those are also poison." Peter Gilila (February 2019:88) added: "Those are definitely proscribed. They should not be eaten when one isn't a raven."

Tulukaruut neqait (baneberries).
KEVIN JERNIGAN

ATSAT
BERRIES

֎

Many types of berries grow in southwest Alaska, and they are harvested in abundance into the present day. The most widespread are salmonberries or cloudberries (*Rubus chamaemorus*) known as *atsat*, *atsalugpiat*, and *naunrat*; blueberries (*Vaccinium uliginosum*) known as *curat*; blackberries or crowberries (*Empetrum nigrum*) known as *paunrat* or *tan'gerpiit*; and low bush cranberries or "red berries" (*Vaccinium vitis-idaea*) known as *kavirlit*, *tumaglit*, or *kitngit*, and discussed above with medicinal plants. Other species include highbush cranberries (*Viburnum edule*) known as *kitngigpiit*, *atsaangruyiit*, and *mercuullugpiit*; northern black currants (*Ribes hudsonianum*) known as *atsaanglluut*; bunchberries or air berries (*Cornus suecica* and *Cornus canadensis*) known as *cingqullektaat*; bearberries (*Arctous alpina*) known as *kavliit* or *kavlagpiit*; northern red currants (*Ribes triste*) known as *agautat*; dwarf nagoonberries (*Rubus arcticus*) known as *puyuraarat*, *puyurnit*, or *puyuraat*; and bog cranberries (*Oxycoccus microcarpus*) known as *quunarliaraat* or *uingiaraat*.

A number of factors are said to contribute to a good berry harvest. Peter Jacobs (October 2003:103) noted that when layers of ice form over snow in winter, salmonberries will be abundant the following summer. John Phillip (October 2003:103) explained: "Snow itself can't stay on the ground for a long time, but if it periodically gets wet [and freezes], it will help the salmonberries when it's time for them to grow. . . . If there is continuous south wind and snow, the snow would stay longer during spring and getting close to summer." John added that rain also promotes berry growth: "Water makes salmonberries grow on the land. They say that when it rains, it gives the land water to drink."

Salmonberries were among the first to ripen, and people paid particular attention to indications that they were ready to pick. Peter Jacobs (October 2003:59) noted: "They always use wild chamomile to predict the occurrence of salmonberries. When the wild chamomiles are large, they say that the salmonberries are ripe." As noted above, an abundance of white cottongrass was also used as an indicator that salmonberries were ripe.

Many elders vividly describe their experiences picking berries when they were young. Edward Hooper (March 2007:1341) of Tununak emphasized how hard women worked: "When I observed my mother and others, they'd go berry picking the whole day, traveling a great distance out there. In the evening, they returned home to this village [Tununak], exhausted . . . because [berries] aren't close by. That's going berry picking using human power." Tommy Hooper added: "These days, they use Hondas [all-terrain vehicles] to go and pick berries."

In the past, berries were ready to pick during August along the coast around Nelson Island and by the end of July in the interior. Today, berries are ripening earlier all across southwest Alaska. Tommy Hooper (March 2007:1335) recalled:

> When it was time, people picked salmonberries. They want to have them for winter so they can have good food to eat. . . .
>
> The area around Nelson Island seems cooler, and [berries] are ready for picking later. But inland in the marshy area, along the low ground, they are ready for picking earlier.
>
> We used to head inland [to pick berries] around August 15. But these days, if one goes during that time, they will be overripe and rotted and no longer good for picking. They are starting to head up there before that time now.

Sometimes berries are hard to find. This absence may be attributed to a number of factors, not all of which are environmental. Nick Andrew (October 2003:212) noted: "They say that when someone is going to die, [berries] won't be seen, even though they are abundant." Nick's wife, Nastasia, elaborated: "Even though there are a lot of berries, when we go to pick berries, there are none where we go. . . . We went berry picking a few months ago and we kept checking for blueberries, but didn't find anything at all. The next day we heard that poor Issaluq had died. [We didn't find berries] because my poor cousin was going to die."

In many areas, when a girl had her first menstruation, she followed *eyagyarat* (abstinence practices), including the instruction not to pick berries for a period of time. Marie Alexie (June 2016:171) noted that when it was time for her to pick berries again, they had her tie a berry down with a strand of her hair. Marie also refrained from picking berries for one year after her daughter died. Teresa Alexie (April 2008:202) of Kalskag shared her experience: "After I abstained for one year, I went picking with my grandma in July for blueberries. She let me take one hair and tie it on the berries and she said, '*Waten piluten, "Amllernek paivuskia"* [Do this, and say, "Make a lot of [berries] available to me"].' So I listened. Gee, lots of blueberries!" Teresa continues to pass this instruction on to young people today.

Alice Mark (October 2009:135) of Quinhagak recalled another important berry-picking instruction: "They say if one comes upon berries very close together, take just one, as those are berries that have been picked by *ircenrraat* [other-than-human persons]." Alice described her experience:

> We saw the berries picked by *ircenrraat* one of the many times we went berry picking. Sammy quickly came over to us, and he said that he saw berries that were very close together over there. Then they started to say, "Maybe these are berries picked by *ircenrraat*." Then we all went over to look at them. We really lacked berries to pick at the time.
>
> Then I recalled that instruction to take just one [berry]. When we went to the [berries] Sammy had found, they were gathered and appeared like someone had put them there, berries that were gathered close together.
>
> I think they were as much as the contents of a cup. They looked like a person just put them there [after picking them]. I finally believed when I saw them.
>
> Since I heard about it, I told the others with me, after first taking one myself and putting it inside my bucket, "Now, each of you take just one." They each took just one.
>
> Then [one person] took all the ones that were gathered together. [*laughter*]
>
> Then after there had been hardly any berries, there were many berries. And our [buckets] were filled right away. We got many berries. I hear that [*ircenrraat*] give [people] berries. They say we should take just one.
>
> We picked berries and [the one who picked lots] was just walking

around up there. When he came, he said that his bucket didn't get filled. [*laughs*]

[He] took all the [berries] that were gathered together. We laughed for a long time. His bucket was empty.

I was thinking, "That [teaching] is evidently true."

Atsalugpiat wall' naunrat
Salmonberries

In coastal areas, salmonberries (*Rubus chamaemorus*) are the most abundant fruit. Known simply as *atsat* (lit., "berries") or *atsalugpiat* (lit., "genuine berries") in some areas and as *naunrat* (lit., "plants") in others, special rules surrounded their use. Wassillie B. Evan (March 2004:363) noted that after a woman gave birth, she was not to eat *akutaq* mixed with salmonberries lest it stop the small openings of her breast: "So if one eats *akutaq* like that while breast-feeding a child—you know salmonberries have many small seeds— those will prevent breasts from filling up. It causes the breasts to suddenly dry up."

Just as Elsie Tommy (June 1992:28) had noted the use of *ayuq* by healers who had gained their powers through insects that stayed on *ayuq* plants (page 115), Elsie said that healers who had gained their powers by placing their hands over insects that stayed on salmonberry plants used salmon-berry leaves:

> You know those insects that aren't furry, the light-colored [caterpillars] with many legs. They say if someone has those kinds of hands, they would use salmonberry leaves [to help them heal people]. If they prepare to use [their hands] to work on someone's body, they [wiped their hands] with salmonberry leaves. And two [leaves] would be used to pry off [the hands] [after using the hands to heal a person's body]. (Fienup-Riordan and Rearden 2017:336)

Theresa Abraham (July 2007:237–39) shared her recollections of accompanying her mother to the low-lying tundra north of Chefornak to pick salmonberries. Wearing waterproof skin boots, they traveled inland and picked all day long. Theresa's mother carried Theresa's younger sibling on her back, using her old hooded garment as a carrying device. Theresa recalled: "The place where we went seemed to always have salmonberries; it would be very

Atsalugpiat (salmonberries), also called *naunrat* on the coast. KEVIN JERNIGAN

red. I would hold a small metal bucket, and my mother also held [a bucket], and she also had an *issran* [twined grass carrying bag]. And when we reached those [berries], I'd look back and see our village a great distance away."

After eating a small meal, the baby fell asleep and Theresa's mother placed it on the tundra, staking something near the baby's face and covering it with her hooded garment while the baby slept. Theresa helped her mother pick, filling her small bucket and emptying it into her mother's *issran*:

And when it was time for us to return home, when she opened her *iss-ran*, I would see that the inner lining [made of dried beluga stomach] was filled [with berries]. She'd carry it on her back, and the straps of those carrying bags were actually very painful during those days. . . .

When she'd place the *issran* on her back, she'd also strap that child along the front of her body . . . and the bucket she was holding was also filled with salmonberries. . . .

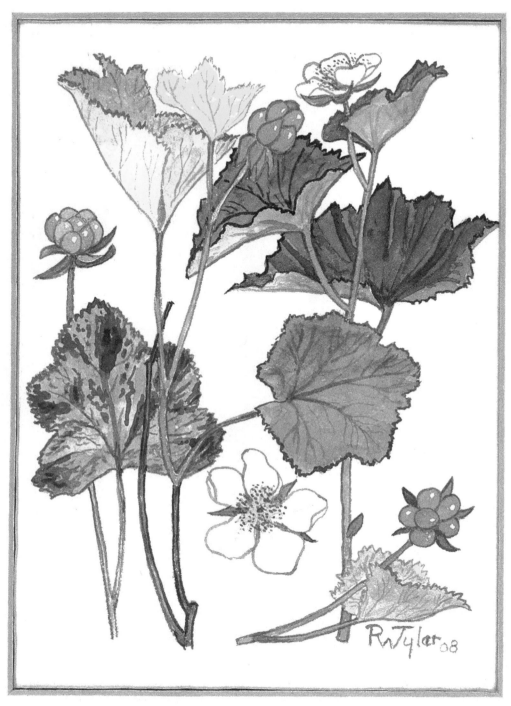

Salmonberries. RICHARD W. TYLER

Fannie Cleveland Moore sitting in a patch of *ayuq* and harvesting salmonberries. JACQUELINE CLEVELAND

That's how she used to take me out to the wilderness. She would bring that child along and not leave [her child] behind.

On the coast, women picking salmonberries used *amrayaat* (carrying bag liners), what Tommy Hooper described as "salmonberry picker bags," made out of a dried seal stomach or beluga stomach and carried inside an *issran*. Martina John (November 2007:37) confirmed that *amrayaat* were only used to pick salmonberries. Susie Angaiak (March 2007:1338) of Tununak recalled: "When I once used a seal stomach for an *amrayaaq*, when my bucket would fill up [with berries], I would empty it out through its opening, and it took me a while to fill. But their contents are good because they are airtight."

Neva Rivers (March 2004:141) also described using seal-stomach containers when berry picking: "We used those stomachs called *qemrayautet*. We never had buckets. They split the bearded seal stomachs that had been inflated, sewed them, and made them rounded. Those can hold a lot. We would pour the berries we picked from those into wooden buckets."

Neither *amrayaat* or *qemrayautet* are placed underwater. To store salmonberries on Nelson Island, women made *naparcilluut* (twined grass storage bags). Albertina Dull (November 2007:37) recalled:

They used to store salmonberries inside twined grass storage bags. . . . They constantly lined the insides [of the grass containers] with sour dock plants, and filled them with salmonberries. . . . They placed sour dock along the outside [of the berries]. And when they reached the opening, they placed sour dock around it and fastened it, using woven grass to fasten it. Then along their openings and along the bottom, along both sides, they placed some wood, and sunk them inside lakes, staking them in the ground.

They tied [the grass storage container filled with berries] onto two pieces of wood that weren't wide and weren't thick, and then they staked them in the ground and sunk [the container] in the lake. And when it froze and was no longer dangerous, when they'd go and get them and bring them indoors, some of the berries were in really good condition.

Anna Pete (December 2012:109) described using the same storage method at the mouth of the Yukon: "[They would twine grass containers] like those [grass carrying bags] . . . but they made them tall. And before they started getting plastic, they would sew the top with king salmon skin and cover them when it was full. Then they sunk them in a lake. Only the outside [of its contents] was a little frozen in the fall. They removed them when it started to get cold out."

John Andrew (April 2019:272) also recalled seeing salmonberries being stored underwater in grass containers:

When they were picking *naunrat* [salmonberries], they would make grass bags from *kelugkat* [water sedge]. Then after packing them tightly, they would have the young men bring [the grass containers full of berries] down to bogs along the side of lakes and have them sink them there, and make a tripod near them to mark it. . . .

Then at freeze-up they said their men would use some kind of small homemade sled to get them. They would have the people who picked berries gather, and they'd saw them into pieces and divide them up to the people who picked them. They'd give even children a share because of their parents.

Anna Pete (December 2012:110) noted that cranberries could be used to cover salmonberries and keep them in good condition when stored in wooden barrels. Angela Hunt added: "When they started to store them inside barrels, a woman named Anna Okitkun, after placing them inside a barrel, she topped the salmonberries with sugar. They didn't get moldy. I think some people also covered them with lard." Anna Pete continued: "Before they got freezers, they tried different [storage] methods. They would also cover them with uncooked sour dock. When I saw one [bucket of salmonberries] covered with cranberries, the [salmonberries] underneath . . . looked like newly picked ones. And they had no mold on them at all."

Curat
Blueberries

Curat (blueberries, *Vaccinium uliginosum*) are another ubiquitous fruit. Ruth Jimmie (April 2019:144) declared: "When I eat these in the wilderness, I can't stop." Bob Aloysius (October 2010:134) recalled picking blueberries along the middle Kuskokwim and turning his mouth blue:

> When it was time to pick berries in the summer, we used to really be eager. We'd take a break from fishing during the end of July, and we'd go and pick berries for maybe one week. They'd go up to the marshlands, going to where the salmonberries had grown, or going to places where blueberries usually grew.
>
> Sometimes, since there aren't many salmonberries in our area, you mentioned that [the land along the coast] looks extremely red [from an abundance of salmonberries]. But [in our area] sometimes when blueberries grew, you could just sit in one spot and pick.
>
> We'd really enjoy eating and would be very blue when we'd eat berries and not pick, going against our admonishment. [*laughs*]
>
> But they really work hard at picking salmonberries and blueberries.

Finally, Katie Jenkins (June 2016:253) shared the well-known tale of how Crane got blue eyes. Trying to improve his vision, he first tried crowberries, then cranberries, and finally blueberries: "After putting on those eyes, he looked around, and gee, they say what he saw was very good. [*laughter*] Those two became his eyes."

Curat (blueberries). KEVIN JERNIGAN

Harley Sundown's grandson with a "blueberry kabob," August 2017. JACQUELINE CLEVELAND

⌃ Harley Sundown's granddaughter holding another ingenious "blueberry kabob,"
August 2017. JACQUELINE CLEVELAND

➤ Misty Sundown with her bucket of blues, August 2017. JACQUELINE CLEVELAND

Paunrat (crowberries). JACQUELINE CLEVELAND

Paunrat wall' tan'gerpiit
Crowberries

Crowberries or blackberries (*Empetrum nigrum*), known as *paunrat* or *tan'ger-piit* (lit., "black ones"), are abundant in tundra hills throughout southwest Alaska. Theresa Abraham (July 2007:239) recalled picking crowberries with her mother near the old village of Cevv'arneq and storing them underwater in a *naparcilluk* (twined grass storage basket):

> One time I told my mother, "The berries we picked." Although I didn't pick berries too well, I was concerned that the berries we picked would spoil. She said that they wouldn't spoil, that I would see them in their original state and not scattered. . . .
>
> And during the fall, she pulled it out of the water, just when the lake was about to freeze. I looked at them and saw that the contents were good. They weren't light in color at all, and they weren't scattered everywhere.

Crowberries were not only harvested for future use. Sometimes jokingly referred to as the La Croix of the north, crowberries could quench a traveler's thirst. John Andrew (October 2019:20) explained: "In mountains far back there, [berries] grow in great abundance. And crowberries, when we'd walk at night, when we'd [come upon] a [large] patch, the ground would make crunching noises. We would squat and feel around for some and satisfy our thirst with their juice."

Mercuullugpiit wall' kitngigpiit
Highbush cranberries

Highbush cranberries (*Viburnum edule*) are known by a number of Yup'ik names, including *kitngigpiit* (lit., "big *kitngit* [low bush cranberries]"), *atsaangruyiit* (from *atsat*, "berries"), and *mercuullugpiit*. Smiling, John Andrew (April 2019:150) recalled the name his coastal wife had given both *tuutaruat* (rose hips) and highbush cranberries because of how she harvested them: "My late wife used to call [highbush cranberries] *nangengqauryarat* [ones that require standing (when picking)]. We call them *mercuullugpiit* [lit., 'something big to drink water with'] in my area, because when they get ripe they get watery."

Mercuullugpiit (highbush cranberries). KEVIN JERNIGAN

Cingqullektaat (bunchberries). KEVIN JERNIGAN

Cingqullektaat
Bunchberries

Bunchberries or air berries (*Cornus suecica* and *Cornus canadensis*) are known as *cingqullektaat* (from *cingqur-*, "to make loud popping noises"). These berries are not plentiful and are not usually gathered in great numbers. John Andrew (April 2019:146) commented: "Their insides are sweet and pop easily. Sometimes on small hills, there are a lot. When we bring kids [berry picking], they like to pick tarty poppers. But they immediately make them into *akutaq* when they don't eat them right away." Theresa Abraham (April 2019:146) noted that although she doesn't make bunchberries into *akutaq* herself, her granddaughter used them to make *akutaq* to share with guests: "These *cingqullektaat*, Angilan wanted to pick them. After expressing how delicious they would be, she said that she was going to pick berries that are sour so that when guests arrive they could eat them and find them delicious. Gee, they had *akutaq*, and left only a little bit uneaten."

Kavliit (bearberries), known to be full of gratitude for those who harvest them. JACQUELINE CLEVELAND

Kavliit wall' kavlagpiit
Bearberries

Bearberries (*Arctous alpina*) are known as *kavliit* or *kavlagpiit*. Like bunchberries, they are not plentiful. Ruth Jimmie (April 2019:156) remarked: "When they make fresh salmonberry *akutaq*, they add those to it. Even though they aren't appetizing, if you eat *akutaq* and you happen to [bite] on one, it is very good." John Andrew added: "People in my area pick them. They say they are raven food or crane food. Before they become ripe, they are sour. Then when they are overripe, they are semi-sweet. And they are good when added to other [berries]. You can also make them into jam when you pick a large amount."

Lizzie Chimiugak (May 2019:173) shared the memorable instruction: "Those elders had *qaneryarat* [words of wisdom] about those *kavliit*. They said that these *kavliit* are grateful by nature for those who pick them. When we pick berries, we remember that those *kavliit* are grateful." Cecelia Andrews

(April 2019:156) also noted that *kavliit* feel gratitude toward those who eat them: "They are grateful to those who like them. They are very thankful. There is a saying that when you show respect to them, you make them happy." Ruth Jimmie added with a laugh: "Probably because they don't taste good." Ruth concluded: "That's why after I found that out, when I see those, after picking them, I eat them."

Quunarliaraat wall' uingiaraat
Bog cranberries

Bog cranberries (*Oxycoccus microcarpus*) are known as both *quunarliaraat* (from *quunarqe-*, "to taste sour") and *uingiaraat* (lit., "husbandless females"). John Andrew (April 2019:148) said that bog cranberries are plentiful in old lakes or floating in wetland areas behind his village: "They are plentiful along wet moss." Theresa Abraham added that the same was true along the coast: "We call them *uingiaraat*, . . . the other kind besides *tumaglit* [low bush cranberries] that grow on moss and that have little *uskurat* [tethers, harnesses]." In fact, another name for bog cranberries along Norton Sound is *uskurtuliarat* (lit., "ones with *uskurat* [tethers]") (Jacobson 2012:690).

Quunarliaraat (bog cranberries), also known as *uingiaraat*. KEVIN JERNIGAN

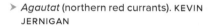
Atsaanglluut (northern black currants).
DENNIS RONSSE.

Agautat (northern red currants). KEVIN
JERNIGAN

Atsaanglluut agautat-llu
Northern black currants and northern red currants

Several other varieties of berries occur in southwest Alaska, including two species of currants. Northern black currants (*Ribes hudsonianum*) are known as *atsaanglluut* (from *atsat*, "berries") or *atsangayiit*. John Andrew (April 2019:152) remarked: "They really smell strong. You can smell them quite a ways off. They are not good tasting. They are hard to eat, but they make them into jam."

Northern red currants (*Ribes triste*) are known as *agautat* (lit., "hanging things"), probably because of the way each berry hangs downward from its stem. *Agautat* are rare along the coast, but are more common upriver. John Andrew (April 2019:152) noted: "They are sour. They also grow abundantly in my home area. . . . They can make them into *akutaq*, and into jam or jelly."

Northern red currants. RICHARD W. TYLER

▲ *Puyurnit* (dwarf nagoonberries).
KEVIN JERNIGAN

▼ *Puyurnit* (dwarf nagoonberries).
KEVIN JERNIGAN

Puyurnit
Dwarf nagoonberries

Dwarf nagoonberries (*Rubus arcticus*) known as *puyuraarat*, *puyurnit*, or *puyu-raat* grow in moist places and are enjoyed as a tasty snack by those traveling in the wilderness. Angela Hunt (December 2012:108) described harvesting this tiny but delicious fruit:

> My husband likes to pick salmonberries right away, once they open up. . . . Then we search for nagoonberries. One time that person made *akutaq* with salmonberries, blackberries along with nagoonberries, mixing everything. Starting with the time he enjoyed eating that, we started picking nagoonberries. All summer they were hard to find. We only found a few just to go along with the *akutaq*.
>
> Some people that I gave [*akutaq*] to during potlatch time, they'd tell me, "This person has such good *akutaq*." Those nagoonberries make it delicious. I never used to pick nagoonberries. Since my husband likes them, I started picking them.

Susie Carl happy for what she harvested from a vole cache near Toksook Bay, on Nelson Island, October 13, 2017.
JACQUELINE CLEVELAND

UUGNARAAT NEQAUTAIT
MOUSE FOODS

❋

Voles and lemmings gather a variety of roots and tubers, which they cache in shallow pockets on the tundra for winter use. Commonly referred to as "mouse food" in English, they can be harvested in the fall by feeling the unfrozen tundra with one's feet. Many described what they called *pakiss-aagluteng* (prying the ground around vole caches to gather tubers). Some of these tasty morsels are eaten raw, while others are highly prized as additions to soups and *akutaq*. They include *anlleret* (tall cottongrass tubers), *negaasget* (silverweed tubers), *utngungssaraat* (grass seeds), *qet'get* or *qetgeret* (horse-tail root nodules), and *elagat* (tubers of alpine sweet vetch), also known as *qerqat*.

Voles have a place in Yup'ik oral tradition disproportionate to their size. In some stories, it was Vole who carried Ingrill'er (Kusilvak Mountain) on his back before dropping it at its present location at the mouth of the Yukon delta. Alma Keyes (February 1993) of Kotlik described how the mouse-like creature *uilurusak* was said to lick every aperture of a person's body, giving that person a long life (Fienup-Riordan 1996:178). Voles are also recognized as dexterous, and a young girl might wear a bracelet made of vole hands so that she would become a skilled seamstress. Raphael Jimmy (January 2013:340) remarked: "And voles, gee, what skilled hands they have. They are actually small things. But when [a vole gathers food], its food cache would have a great many of those [tubers]."

Martina John (November 2007:1) described how voles could be used as indicators of the approach of cold weather: "When the voles hadn't gathered food in their caches, they knew through those voles that winter wasn't going to come [early]; they knew that it wouldn't get cold quickly. And when they started to have food [in their caches], they knew that it was going to be winter." Martina's brother, Jobe Abraham (February 2019:78), also described how people valued vole feces as medicine: "It seems they crushed their feces and put them in water and let people who were dehydrated use them."

The role of voles as food providers is also highly valued. Theresa Abraham (April 2019:126) declared: "The foods [voles] gather are amazing." David Martin (May 2004:16) recalled: "Those voles that don't seem important to us would gather food for people's stomachs. They would gather the roots of plants from underground." Frank Andrew (February 2003:833) noted: "There are vole caches down in our village, [with food] including *anlleret* [tall cottongrass tubers], *negaasget* [silverweed tubers], *utngungssaraat* [grass seeds], *agivaat* [English name unknown]. But here on the Kuskokwim, they call them *agivaat naunrat*. The voles allow us to quickly gather that food." Neva Rivers (May 2004:76) said: "[People] don't gather the tops above ground, but when the voles get their roots, we get *utngungssaraat* and *marallaat* [silverweed tubers]."

Albertina Dull (November 2007:1) added that not all the foods that voles gather are edible, such as the tubers of poison water hemlock: "They have various types of food, but we don't eat some of their food because they are inedible." Helen Walter queried: "I wonder why some [voles] gather food that is inedible, or is it their dessert?" Albertina (May 2019:169) specifically mentioned *cuqlamcaat* (the edible tubers of pink plumes, *Bistorta officinalis*) found near Up'nerkillermiut: "They are unpalatable. I don't think people gathered those, but they're food."

Martha Mark (January 2011:251) described gathering a variety of tubers from vole caches along the shore below Quinhagak:

> Long ago, when my grandmother and I used to go, from that place down the coast, from underneath we gathered *negaasget* [silverweed tubers]. Quinhagak has always had *negaasget*. They are nice, and they bloom yellow flowers. Their tubers aren't long.
>
> And those *aatuuqerrayiit* [English name unknown] look like small Christmas trees. Those grow down along the ocean shore and around Ceturrnaq River. Nayak'aq gives me lots of those. This past fall, she gave

me many and had washed them. They are some sort of [mouse food] that are mixed in with *utngungssaraat* [grass seeds]. When we go in search of *anlleret*, they are mixed with them, but there aren't many of them.

Qet'get [horsetail root nodules] aren't around the marshland, but are situated among trees. They look like blackberries. They are also delicious and taste sugary when made into *akutaq*.

And there are many *utngungssaraat*. Sometimes when prying the ground [where the vole cache is], it is only filled with *utngungssaraat*. They are especially around the mouth of Ceturrnaq River; we used to get them from that area back when we went and gathered food from the land.

Speaking about gathering mouse food, John Phillip (October 2010:130) recalled: "Those small [voles] sometimes make large [caches], and they even make one beyond [the other]. It's fun going in search of those, prying the ground. Sometimes one of those [carrying baskets made of grass] is filled from just one vole cache." John (February 2005:134) also noted: "These [voles] that make you squeamish gather food for us. . . . They apparently harvest those *negaasget* for us. . . . And they also gather the tubers of *iitaat* [tall cottongrass] that are along the edge of lakes. They gather those *anlleret* out on the coast to make *akutaq*; those taste like sugar."

Mary Napoka (May 1989:33) spoke about harvesting mouse food upriver from Bethel: "*Anlleret* are food and *elagat* also and *qet'get* [root nodules of horsetails], and they call those things that are small and green *aatuuqerra-yiit*. They say that the entire land is food." Bob Aloysius (October 2010:125) added detail for the area around Aniak:

In the fall, when it began to get cold, they would bring us to islands. . . . While running along the shore sometimes, our legs would suddenly sink in. They'd suddenly get happy and quickly come to us, and they'd pry open the ground, and there would be many things, and even *qet'get* with them. Those things that the poor voles worked so hard to gather, we'd steal them. [*laughter*]

They'd tell us to replace it with something small, and even a small piece of dry fish, showing gratitude for voles gathering those.

Paul John (May 2004:76) noted that weasels also store food: "They say those *teriaraat* [weasels, ermines] have [caches filled with tubers]. They say those little weasels eat *iitaat*. Those are the ones that have very clean stored food." David Martin elaborated: "The ones that the *teriaraat* store in caches

△ Susie Carl uncovering a vole cache of *anlleret* (tall cottongrass tubers) mixed with sticks
   and other inedible tubers. JACQUELINE CLEVELAND

▽ Susie Carl holding three *anlleret* (tall cottongrass tubers) from the cache. JACQUELINE CLEVELAND

are cleaner than the ones voles store away. They look as though they've been stored in containers. Voles are messier than these *teriaraat*. The *teriaraat* cleaned them very thoroughly, they were very meticulously [cleaned]." Paul John made a distinction between *teriaraat* and *narullgit*, both translated as weasels in the Yup'ik dictionary (Jacobson 2012:429, 638), and *puveltut* (red-backed voles): "The small ones without tails are *teriaraat*. . . . They call the other type *puveltut*, and they would say they are *ellamiutat* [lit., 'ones from the sky']." Neva Rivers gave a similar name for lemmings: "They would say they are *qilagmiutat* [lemmings, lit., 'ones from the sky']. They say when those gather food, they are very clean."

The saying that lemmings and voles come from the sky may refer to the fact that they can seemingly appear out of nowhere in large numbers. Cecelia Andrews (April 2019:314) recalled her older brother's son's experience when spending a night in the wilderness. Climbing inside a boat to sleep, he and his companions found the boat filled with *puveltut*: "And he said they could not spend the night at all. Even though they were tired, . . . they returned home. On their trail, they said as they were walking, there were more [red-backed voles] walking all over. And then someone said that they had fallen [from the sky]."

Many people are creeped out by voles and shrews, but not John Andrew (April 2019:312):

> When I am in the wilderness, even though they were on my body, I didn't suddenly shudder. . . .
> When I got tired, I laid down on my sleeping bag and stayed there. I saw one come out of my pocket. [*laughter*]
> And when my other uncle was with me, [he'd tell me], "Gee, you're not creeped out at all. Those are moving on your body." I told him, "They won't hurt me, they're small." He's actually a big person; when they would scurry over his body, he would suddenly jerk.

Finally, some note that voles are like people: some are lazy and some are clean. John Andrew (April 2019:177) observed: "After looking at [vole caches] that were filled with all sorts of things, I would sometimes cover them again. I would think, 'Those who are lazy probably gather a mixture of things.' . . . But some of them had gathered very clean [food] that were one type [of food]." Cecelia Andrews (April 2019:177) reported that mice see different types of tubers as different seals: "They said *utngungssaraat* were ringed seals. And they said those *marallat*, the long, small potatoes, they'd call them *makliit* [adult bearded seals]. And they said the ones that were the most unappetizing, they called them *tegat* [ringed seals in rut, smelly and inedible] . . . . They also say [voles] are *ircenrraat* [other-than-human persons]."

Theresa Abraham (April 2019:179) commented: "Like us people, we don't mix our foods. When this type of food is different, we put them in separate containers. Since they know about them, they probably divide those foods like we do." Ruth Jimmie added: "Indeed. When they tell their children, 'Go and get *utngungssaraat*,' they'd know what they are." Cecelia Andrews added: "I used to hear that when those mice prepared food close to winter, they would tell them to gather [foods]. They said when they were lazy, . . . and when they wouldn't listen, they would also [scold them]. . . . They said they would gather around him and really scold that one, telling him not to be lazy." "Like people," Ruth said.

Anlleret
Tall cottongrass tubers

Among the most commonly gathered tubers are those of *iitaat* (tall cottongrass), widely referred to as *anlleret*. Wassillie B. Evan (December 1985:101) recalled: "They would eat *anlleret* with fish eggs. They would mix them with salmon eggs. And when they didn't eat them like that, they would cook them. They would dice them up, then mix them into *akutaq*, adding no other berries to the mixture. They were delicious."

Raphael Jimmy (January 2013:341) described how his mother prepared *anlleret*: "Sometimes she would cut them up first after washing them. . . . When my mother was going to make those into *akutaq*, she used to cook those. When they didn't cook them, we would eat them with seal oil." Francis Charlie (November 2014:169) noted that he preferred them added to *akutaq* uncooked.

Fall was not the only time that people gathered mouse food. David Martin (May 2004:98) described finding *anlleret* in spring that a vole had pushed out of its cache in an area where the snow had melted from the ground:

> They appeared like a group of ptarmigan pellets that were partially melted, and there were many of them before me. I took one and broke it in half. It was *anlleret* that [voles] had gathered in a cache. I took some, and I went to my mother and showed her. The one whom I went to quickly got ready, "Let's go get them right away." . . . She filled that large woven grass basket and would say, "Oh my, you found a lot of food to eat."

A stalk of tall cottongrass, showing the tuber that a mouse might harvest. KEVIN JERNIGAN

> That's what happened to me back when we were going through a food shortage. . . . As we are living, the unseen entity causes us to go through hardship so we can come to our senses.

Neva Rivers (May 2004:48) added: "We searched for food that the voles took out of their caches, and we'd gather a large amount. They take out their food to the surface of the snow from underground caches before water fills them. We eat those, emptying their caches of food. When we came upon those, we never left them alone and brought them home during spring."

Martha Mark (January 2011:258) had also gathered *anlleret* in spring: "When we'd travel in the spring, we'd find some [*anlleret*] that [voles] had pushed outside of their homes. When they weren't dry, we'd take them, and when they were dry, we'd leave them." George Pleasant added that vole caches were filled with *anlleret* in places with a lot of tall cottongrass.

Like all living things, *anlleret* were treated with care. Alexie Nicholai (April 2015:315) recalled the admonishment never to carry *anlleret* across the upriver side of a fish fence in the fall lest they startle the fish: "These fish are apparently easily offended. . . . They didn't bring things across on the upriver side, but only downriver."

Negaasget utngungssaraat-llu
Silverweed tubers and grass seeds

Vole caches along the coast contain both *negaasget* (tubers of Eged's silverweed, *Potentilla anserina* subsp. *groenlandica*) as well as *utngungssaraat* (grass seeds, from *utnguk*, "wart"), which some elders say are specifically from *taperrnat* (rye grass). John Phillip (February 2005:134) described them as the "potatoes" and "onions" of voles: "Those [voles] have good food. They have potatoes and onions. That is what I call the *negaasget* since their surface resembles potatoes and their insides also look like potatoes. And I call *utngungssaraat* the onions of voles. That's what they harvest." Frank Andrew (February 2003:833) agreed: "Those *negaasget* are potatoes. That's what they are when they are cooked, potatoes. They are good to add to food." John Andrew (April 2019:135) recalled the time that John Phillip introduced him to these tasty morsels: "When they gave me some [*utngungssaraat*], John Phillip, an elder at Kongiganak, said to me, 'What do you think those are?' When I replied, they started laughing. I told them they were nothing to me, that they were debris. When I tried them; they were delicious."

Lena Atti (February 2005:28) noted: "The [plants] that have *negaasget* tubers grow in summer, with yellow flowers on them in marshy areas among *evesraat* [soft grasses]." Many people like adding *negaasget* to soups, and they are also good in *akutaq*. Pauline Jimmie (March 2007:71) said that both *utngungssaraat* and *negaasget* are delicious when added to broth soup and seal meat soup. Ruth Jimmie (April 2019:57) described mixing *utngungssaraat* with crushed tomcod livers, tomcod eggs, and seal oil: "They are very delicious. They call them *uuqnat*."

Cecelia Andrews (April 2019:175) mentioned finding *neqnialqut* (lit., "bad tasting things") along with *utngungssaraat* when searching for mouse food along the coast: "Those *neqnialqut*, when prying the ground for mouse food. . . . there are some *utngungssaraat* mixed in, and then roots that are dark; eating them is very unappetizing. They don't taste good. They also remove them and take those *utngungssaraat*." Theresa Abraham agreed: "We pick the *negaasget* and *utngungssaraat* down on the marshland, and we leave those that don't taste good alone."

A vole cache containing a mixture of *negaasget* (silverweed tubers) and *utngungssaraat* (grass seeds).
JACQUELINE CLEVELAND

Grass seeds and silverweed tubers cleaned after the harvest.
JACQUELINE CLEVELAND

⌃ A handful of *negaasget* (silverweed tubers). JACQUELINE CLEVELAND
⌄ Teardrop shaped *utngungssaraat* (grass seeds). JACQUELINE CLEVELAND

Qetgeret wall' qet'get
Horsetail root nodules

Qetgeret, also *qet'get*, are the root nodules of *tayarulunguat* (horsetail plants, lit., "fake mare's tail," *Equisetum* spp.). Also known as "waterberries," they are gathered by voles. Francis Charlie (January 2013:340) recalled: "[*Qet'get*] are delicious. They are long and have things on them that resemble crowberries. And they are white on the inside." Edward Adams (November 2014:169) of Nunam Iqua agreed: "They call those *qet'get*. They look similar to crowberries." John Andrew (April 2019:65) told how his sister prepared them: "I said to my older sister, 'What are you washing?' After laughing, she said that she was washing the *qet'get* in the washing machine. . . . The old people liked to make them into *akutaq* while they were raw. Then some people cooked them a little, but not boiling them for a long time." Nick Andrew (February 2019:20) agreed that *get'get* were delicious when added to *akutaq*.

Joe Felix (July 2007:627) described how his grandmother used the juice of *qetgeret* to treat tuberculosis: "When her son Cakuucin's health started to deteriorate downriver at Talarun, she dried *qetgeret*. And when they withered, she'd boil them in water. She'd have him drink the juice in the morning and evening when she cared for him." Albertina Dull (May 2019:168) stated that some used *qetgeret* as medicine, but she did not elaborate. Ager and Ager (1980:33) do not mention horsetail tubers, but do report the use of horsetail plants to brew a medicinal tea.

Susie Carl holding *qetgeret* (horsetail root nodules), also known as *qet'get*, along with a horsetail plant. JACQUELINE CLEVELAND

▲ Horsetail plant.
KEVIN JERNIGAN

▼ *Qetgeret* (horsetail root nodules).
KEVIN JERNIGAN

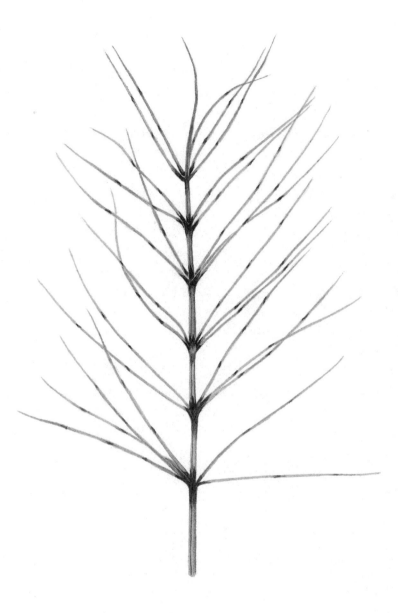

©Sharon Birzer

Horsetail plant and root nodules. SHARON BIRZER

Plant tubers are not the only things found in vole caches. Voles are hoarders, and many tell of finding unusual things in their underground homes. Frank Andrew (February 2003:542) told of an elderly woman from Cal'itmiut who found the tail of a beluga whale in a vole cache: "It was still bleeding. She put it in her grass carrying bag and brought it home. When she got home, she cut it up and cooked it, and they all ate it. She is still living at Kwigillingok." Frank said that another woman found a slab of dried king salmon in a cache: "She brought me what she had found and wanted me to eat some of it. . . . When it was time to eat, I ate the slab of king salmon and it was good. It smelled like good dried fish. . . . Nothing happened after we ate it. These voles are fascinating. I ate dried king salmon that was dug from a vole's cache."

Paul John (May 2004:20) noted that one of his children uncovered an egg in a vole cache while searching for mouse food, while a person from Nightmute uncovered a Canada goose inside a vole cache. John Andrew's sister (April 2019:120) once uncovered a large frog in a vole cache, then yelled and left it alone: "They said if she had taken it and had it as a pet, she would have been rich. And they went to it, but just like [in previous stories], it was gone. They say something displays things to some people. . . . They tell them not to be afraid of them when they uncover those in vole caches."

Neva Rivers (May 2004:21) recalled how people she knew uncovered needlefish that were still moving inside a cache and ate them all, with no ill effects. In fact, she noted that eating such foods is said to give one the power to heal: "If they were to take a piece and eat it, they could heal people with their saliva. . . . My late older sibling Alexis told me that you must have a small piece, that if you eat a small portion of that, just two or one, it will heal [if you rub the wound with your saliva]. Since these aren't explained as they should be, [young people] don't think that what we heard is true."

People may also acquire the power to heal when they uncover small insects or other unusual items in a vole cache. One is told to place both hands over the insects, and the ability to heal will enter one's body. David Martin (May 2004:22) described his experience:

> Since those elders of the past knew, they said that voles display things that bestow healing abilities to some people. . . .
> Since long ago, these voles that seem so insignificant have displayed things that heal people. . . . My late mother, when I was a certain age,

she went out to go search for mouse food from caches during fall before freeze-up and asked me to carry them on my back.

Before that time, I also found a vole cache filled with spiders that are a little larger than maggots that were struggling with one another down inside the cache. Since I didn't know what they were, I covered them and left them.

As I was on my way, [my mother] called me. . . . When I got there, in the center of a vole's cache down there was a hand. It was this [left] side. But this [index finger] had no tip. When I got to her, she said to me, "I think they say these voles display things that help people. Now, you're young, cover those down there with the bottom part of your garment and line your hands up on top of them."

Just as I placed it there, a tingling feeling went up starting from the tip of my fingers. And then when it got to this point, it went away. . . . And when we uncovered it [those bones] were gone. My hands probably felt tingly when the [bones] went inside [my body]. "I think you're probably going to start healing people." That's what my mother said to me.

Later in life, David Martin used his hands to heal his grandmother, by touching her in the area that was painful below her breast. Paul John (May 2004:23) concluded that some people still uncover things in vole caches that give them healing abilities, but many fail to receive this ability since they don't know about it.

QANEMCIT QULIRAT-LLU

STORIES AND TRADITIONAL TALES

Yugtun Igautellrit Kass'atun-llu Mumigtellrit
Yup'ik Transcription and Translation

The Central Alaskan Yup'ik language is spoken on the Bering Sea coast from Norton Sound to the Alaska Peninsula, as well as along the lower Yukon, Kuskokwim, and Nushagak Rivers. It is one of four Yupik languages, all of which are closely related to the Inuit/Iñupiaq languages of the Arctic coast of Alaska, northern Canada, and Greenland, although they are not mutually intelligible. Together, Inuit/Iñupiaq and Yupik constitute the Eskimo branch of the Eskimo-Aleut family of languages. No apostrophe is used when speaking of Yupik languages generally, but an apostrophe is used for Central Alaskan Yup'ik and its dialects.

There are five dialects of Central Yup'ik: Norton Sound, Hooper Bay/Chevak (Cup'ik), Nunivak Island (Cup'ig), Egegik, and General Central Yup'ik. All are mutually intelligible with some phonological and vocabulary differences (Jacobson 2012:35–46; Woodbury 1984:49–63).

The Central Yup'ik language remained unwritten until the end of the nineteenth century, when Russian Orthodox, Moravian, and Jesuit Catholic missionaries, working independently of one another but in consultation with Native converts, developed a variety of orthographies. The orthography used consistently throughout this book is the standard one developed between 1967 and 1972 at the University of Alaska Fairbanks and detailed in works published by the Alaska Native Language Center and others (Reed, Miyaoka, Jacobson, Afcan, and Krauss 1977; Miyaoka and Mather 1979; Jacobson 1995).

The standard orthography for Central Yup'ik represents the language with letters and letter combinations, each corresponding to a distinct sound as follows:

Consonants

	Labials	Apicals	Front velars	Back velars
Stops	p	t, c	k	q
Voiced fricatives	v	l, s/y	g (ug)*	r (ur)
Voiceless fricatives	vv	ll, ss	gg (w)	rr
Voiced nasals	m	n	ng	
Voiceless nasals	m	n	ng	

* Symbols in parentheses represent the sounds made with the lips rounded.

Nelson Island elders Rita Angaiak, Anna Agnus, Martina John, and Joe Felix exploring the tundra along the upper Cakcaaq River, while young people pick berries, August 2007. ANN FIENUP-RIORDAN

Vowels

	Front		Back
High	i		u
Mid		e	
Low		a	

The apostrophe indicates consonant gemination, or doubling (and serves several other less important functions). There are also conventions for undoubling the letters for voiceless fricatives under certain circumstances (Jacobson 1995:6–7). This standard orthography accurately represents the Yup'ik language in that a given word can be written in only one way and a given spelling can be pronounced in only one way. Note that certain predictable features of pronunciation, specifically automatic gemination and rhythmic length, are not explicitly shown in the spelling.

Most of the translations in this book were done by either Alice Rearden or Marie Meade. As translators, Alice and Marie offer distinctive strategies for bridging differences between Yup'ik and English without erasing them. For both, the goal has been a natural-sounding, free translation, as opposed to either literal translation (at one extreme) or paraphrasing (at the other). Paraphrasing may communicate some of the sense of the original, but such interpretive translations modify the original to the point where the speaker's voice is alternately erased or transformed. Literal, word-for-word translation also falls short. At best, it is awkward, and at worst, it makes no sense. The narrator's choice of words is respected in this book, although translators may modify word order and sentence structure slightly to communicate original meaning. They do this in different ways. Marie Meade, for example, is freer with English word choice, paragraphing, and paraphrasing, in contrast to Alice Rearden, who retains a more literal word choice and style. David Chanar, originally from Nelson Island, also contributed several translations for this book, and his playful sense of humor enlivens his work.

Because their primary goal is communication, no translation in this book mechanically follows the structure of the original language. For example, Yup'ik word order is "English turned on its head," in which suffixes indicating tense, person, case, and other units of meaning are appended to verb and noun bases. Thus, the English phrase "my little boat" corresponds to the single Yup'ik word *angyacuarqa*, which consists of *angya-*, "boat," plus *-cuar-*, "little," plus *-qa*, "my," so that the order of the parts within the Yup'ik word is "boat, little, my." In Yup'ik discourse, the object also typically precedes the verb. A literal translation of *qaltarpaliunga* might read "bucket/big one/ to make/I." A more natural translation would employ typical English word order, that is, verb followed by object, and would read "I/make/a big bucket." Thus, translation involves a continuous process of reordering.

Other characteristics of Yup'ik oratory have been carefully retained. For example, redundancies and repetitions are important rhetorical devices in Yup'ik narrative. Narrators frequently restate important points, often phrased somewhat differently, at the beginning and end of an account, both to enhance memory and to add emphasis and depth. Use of repetition gives Yup'ik texts a denser texture than typical English phrasings, which careful attention in the translation can retain. Structured repetitions are characteristic of Yup'ik narrative art and vital to its structural integrity. To smooth them over or omit them would impoverish the translations.

Several grammatical features of the Yup'ik language pose potential problems for translators. First, relatively free word order characterizes the Yup'ik language. For example, the meaning of the English sentence "The man lost the dog" can only be conveyed by placing the words "man," "lost," and "dog" in this order. A Yup'ik speaker, however, can arrange the three words *angutem* ("man"), *tamallrua* ("s/he lost it"), and *qimugta* ("dog") in any of six possible word orders with no significant change in meaning. Nevertheless, word order is not totally irrelevant to interpreting Yup'ik sentences. Word order may be the only key to appropriate interpretation when the ending alone is insufficient. For example, the sentence *Arnam atra nallua* (lit., "woman//his/her name//s/he not knowing it") can mean either "The woman does not know his name" or "He does not know the woman's name." The same three words in a different word order, however, are less ambiguous. *Arnam nallua atra* is commonly taken to mean "The woman does not know his name." In contrast with other languages that have a free word order, the relative position of postbases inside a Yup'ik word is very rigid. Consequently, syntactic problems may occur in words that occur only in sentences in translation.

Translation is further complicated by the fact that the Yup'ik language does not specify gender in third-person endings. The listener is left to deduce gender from the context of the account. When a speaker describes women's tasks, we have translated the pronominal ending as "she," as that is the way an English speaker can best understand the speaker's intent. Conversely, pronominal endings are translated as "he" when the speaker is describing a man's activities. In general discussions, we have used either "it" or "he," depending on the context. Readers should also know that Yup'ik orators sometimes mix singular and plural endings in a single oral "sentence," and we have retained these grammatical variations to reflect the complexity of the Yup'ik original.

Yup'ik verb tenses also differ from English tenses. Although some postbases place an action clearly in the future and others place action definitely in the past, a verb without one of these time-specific postbases may refer to an action that is happening in either the past or the present (Jacobson 1984:22). Accounts of events or customs that are no longer practiced in southwest Alaska have been translated in the past tense. Readers should also note that tense may vary within a paragraph, especially in discussions of *qanruyutet* (oral instructions) marked by the enclitic "-*gguq*," which can be translated "they said," "they say," or "it is said," depending on the context. Traditional *qanruyutet* that speakers indicate still apply are translated using the present tense.

Our narrators also frequently used nonspecific pronouns and phrases that are difficult for English readers to follow. For example, a speaker may say "that one who told the story," rather than naming a specific person. Narrators also often use phrases such as "he went down" or "he arrived" without specific places mentioned. Readers should note that the Yup'ik language has an elaborate set of demonstratives that situate listeners and that indicate relative placement of action and movement of people—often very specifically—without ever mentioning places directly. These include terms such as *pikavet* (toward the area up above), *piavet* (up the slope), *kanavet* (down the slope, toward the area down below), and *uavet* (toward the mouth of a river, toward the door), to name but a few (Jacobson 2012:963–67). Demonstratives also distinguish between things upslope, downslope, and so on, that require more than a single glance to be seen, things that can be seen fully in a single glance, and things that are obscured from view. Where necessary, we have tried to clarify these phrases using brackets to indicate the narrator's intent. We have used parentheses to designate passages where narrators themselves offer explanations important for the reader but not necessarily part of the account.

Many narrators attach the postbase "-*miut*" (people of) to the name of a river or slough to designate the people living there, as in Kusquqvagmiut (the people of the Kusquqvaq [Kuskokwim River]). The names of many villages also derive from the name of the river or lake where they are located, for example, the old village of Luumarvigmiut on the Luumarvik River. However, narrators may also use the name Luumarvik for the village itself, and in fact often do so. Other village names may be rendered with or without the "-*miut*" ending. The maps that accompany this text show the most commonly used place name. The text, however, reflects what narrators actually said, designating the place with or without the "-*miut*" ending.

Yup'ik oral rendering values close attention to detail and consistent retellings, and whatever their stylistic preferences, Marie and Alice continue to work in that tradition. As Yup'ik scholar Elsie Mather (1995:32) notes, "The most respected conveyers of Yup'ik knowledge are those who express things that listeners already know in artful or different ways, offering new expressions of the same."

Igautellrit
Transcription

As if translation from one language to another were not challenging enough, this book involves the movement from oral to written language. Our starting point is the verbal artistry of individual elders, but critical to understanding their words is the transfer of their voices onto the page. Through the 1970s, little attention was given to reflecting the dynamics and dramatic techniques of the performance, including the speakers' shifts in tone and rhythm. The oral origins of texts were all but hidden from view. Texts were routinely transcribed in paragraph form, as if the paragraph were the "natural" form of all speech.

Beginning in the 1980s, when so many basic tenets of anthropology were being scrutinized, the ubiquitous paragraph came under attack, especially in the work of Dell Hymes (1981) and fellow linguist Dennis Tedlock (1983). Together Tedlock and Hymes inspired a generation of linguists and anthropologists who have since adopted and adapted their insights in a variety of sociolinguistic transcription styles, igniting a veritable "renaissance" in the translation of Native American literature (Swann 1994:xxviii). Although neither Alice nor Marie have chosen to employ the "short line" verse format favored by many translators, they use the prose format with a new sensitivity. In their work, paragraphs are no longer arbitrary groupings disconnected from the speaker's original oral performance but are distinguished by prominent line-initial particles like *tua-i-llu* ("so then"), by cohesion between contiguous lines, and by pauses between units. This is by no means a mechanical process, however, and different translators make different choices about what markers require a new paragraph.

As we think about both the limitations and power of translation to communicate meaning across cultural and linguistic boundaries, it is useful to recall that translation is not the endpoint of understanding, but the beginning (Becker 2000:18). Similarly, the reader is invited to engage these translations and use them as starting points for understanding and respecting the profound differences between literary traditions that, in turn, make it possible for us to better understand ourselves.

MAKIRAYARAQ
GATHERING FROM THE LAND

Molly Alexie and her children, Tristan and Nicolette, heading home with bags full of fresh beach greens, June 2019. JACQUELINE CLEVELAND

Before Ellam Yua covered the land with his large hand[5]

Paul John: Now, I just want to briefly mention this example. These people here just talked about gathering grass that would be used before there was snow on the ground.

With an example, our ancestors urged them to work hard to gather [grass] before there was snow on the ground, the grass that would be twined together, grass for boot insoles, ones they would use for everything, before Ellam Yua covered the land with his large hand, *Ellam Yuan unalvallraminek nuna pategpailgaku*, as they say.

Snow covers these grasses that grow. They used an example for their efforts at gathering [grass] before the snow covered them, that [they should gather grass] before Ellam Yua covered the land with his large hand.

Thinking that you hadn't heard the example for that teaching, I just wanted to briefly mention it; the instruction to gather [grasses] before snow covered them, before Ellam Yua covered the land with his large hand, *Ellam Yuan unalvallraminek nuna pategpailgaku,* as they say. The snow was likened to Ellam Yua's hand, if he covered it with snow.

Buttercup plants[6]

Theresa Abraham: They gather *kapuukaraat* [buttercup plants] during the spring when the ice starts to melt. The bottom of the lake has ice on it. Those *kapuukaraat* are visible, and they obtain them.

Alice Rearden: How do they obtain them?

Theresa Abraham: They use a wooden tool with a [crooked] tip like this. It's pretty long, and they said they used it as an *ayaruq* [walking stick]. They said long ago, women always used the walking stick every time they went out to the wilderness. And when we were about to go on lakes, they always had us use a walking stick. They used the [tip of the tool] that looks [bent] like this to gather those [buttercup plants], and the tip along the end was used to check the area where we were about to walk. We always [jabbed] the place where we were about to go down [into the lake] in the springtime.

Ellam Yuan unalvallraminek nuna pategpailgaku

Kangrilnguq: Tua-ll' augna ava-i *example*-aara qanrutkeqernaluku. Ava-i ukut canegnek cumercaraq qanrutkekiit atuugarkanek qanikcanga'arpailgan.

Example -aangqerrluku-am augkut ciuliamta, cumiggluki qanikcangvailgan pitaqengnaqesqumalallruit tamakut atu'urkat can'get tupi'igkat, piinerkat, aturarkat-wa tua-i callermeng tamiini wagg'uq Ellam Yuan unalvallraminek nuna pategpailgaku.

Tua-i-am qanikcaq patutektukiit makut naumalriit can'get. Qanikcaam patuvailgaki pilallerteng tauna taringcetaarucirluki-am, Ellam-gguq Yuan unalvallraminek nuna pategpailgaku.

Tuaten tua-i tamana *example*-aara niiteksaicukluku qanrutkeqernaluku piaqa; aug'um qanikcangqerpailgaki pitaqengnaqsaraat, wagg'u-q tua-i Ellam Yuan unalvallraminek nuna pategpailgaku. Tamaa-i qanikcaq-am una unalvallerqevkarluku Ellam Yuanun, patkaku tua-i qanikcamek.

Kapuukaraat

Paniliar: Up'nerkami kapuukartetuut ciku urugartaqan. Camna acia nanvam cikutarluni tua-i. Tamakut tamaa-i alaunateng kapuukaraat pitaqluki.

Cucuaq: Qaill' pitaqluki?

Paniliar: Aug'umek-am, waten-a muragaq iqua ayuqluni. Tua-ll' takrarluni, ayaruqluku-gguq. Ak'a-gguq tauna tua-i arnat yuilqurrneq-llu tamalkuan ayaruq pegingayuitelaraat. Nanvanun tua-i piqataqumta ayaruicessngaunata-llu. Tauna tua-i waten ayuqelria tamakunek pissuutekluku, kan'a-ll' iqua tua-i tamatum ciunerkamta yuvrircuutekluku. Tua-i waten tua-i tauna atraqatalput piqerrlainarluku ciunerput up'nerkami.

Then along our walking sticks, the section that is like that, the tip that looks like this, we used it to gather *kapuukaraat*, using it to cut [*kapuukaraat*] down [in the lake]. Then we picked those *kapuukaraat* and placed them in our containers.

Martina John: How do they gather them? Do they appear [when they come to the surface]?

Theresa Abraham: They come to the surface and appear when they are cut underwater, since the bottom of the lake is ice. When we go down [in the lake], if the ice won't break, we go down. If the ice down [along the bottom of the lake] is broken, [lakes] are very deep during spring. And if we are able to go down, after sticking our walking stick in the water and searching for them, using the tip of the tool, we cut them under the water along their ends, and then those come to the surface and appear.

Martina John: How, right after it freezes?

Theresa Abraham: During spring, before the ice along the bottom of the lake melts. The lakes out there are very deep when the ice has melted.

They aren't mud, but the bottoms [of lakes] are sort of swampy. Then when those come to the surface, we pick them. But the ice is cold.

We wade in the water.

That's what *kapuukaraat* are like. They cook them. They check them when they cook them. When they are overcooked, they tend to break up into small pieces, like those *negaasget* [silverweed tubers] or *utngungssarat* [grass seeds], they tend to easily disintegrate. They can become overcooked and turn bad. But when they check [the *kapuukaraat*] while cooking them, they stop cooking them when they are just right [not mushy]. Then [having cooked them] they put them in soup or they eat them, even if they aren't placed in soup. They are delicious.

Tuai-ll' tamakut tamaa-i ayaruput, tuaten ayuqelria, pikna waten kangra ayuqelria, nutaan kapuukarnek pitsuutekluku camavet kepurtuutekluku. Tua-i tamakunek avurluta augkunek kapuukarnek avurluta tua-i assigtamtenun piaqluki.

Anguyaluk: Qaillun pilartatki? Puglartut-qaa?

Paniliar: Puglartut, kep'arqelluteng camani, cikuulaami nanvam acia. Atraraqamta tua-i imyugnaitaqan tamana atrarluta. Imumakuni camna, et'ukayalartut up'nerkami. Tua-i-llu atraryugngakumta ayarumtenek kautuarraarluki, nutaan tamakut taumek iquanek waten camavet kepurluki tua-i ngelaitgun, pugluteng tamakut.

Anguyaluk: Qaillun, cikunrakun-qaa?

Paniliar: Up'nerkami, nanvam terr'a camna cikuirpailgan. Et'ukayalartut-am avani nanvat urumaaqameng.

Marayauvkenateng angayaarrluunganateng aciit. Tua-i-ll' tamakut pugleqata tua-i avurturluki. Taugaam kumlataqluni cikuq.

Ivrarluta.

Kapuukaraat-wa tua-i tuaten ayuqelriit. Tua-i kenirluki. Naspaaqerluki tua-i keniquneng. Kenivallakuneng-am caranglluqerrluteng augkucetun negaasegtun wall'u-q' utngungssartun caranglluqercukaarluteng. Tua-i iciw' kenirpallagvigluteng, assiirrluteng. Taugaam naspaaqerluki tua-i keniquni pitacqeggiqerreskan taqluku. Tua-i-ll' *soup*-anun wall'u tuaten *soup*-aunrilengraan nerluki, kenirluki. Neqniqluteng.

Beach greens[7]

Martha Mark: And when the time to harvest buttercups passes, they gather *it'garalget* [beach greens]. We started to pick beach greens back when I used to go.

Beach greens also grow a second time. The first ones that grow get hard, and they grow yellow things on them. Then in July, others grow along their branches. Those are good and soft.

Pauline Matthew: Beach greens grow down along the ocean shore all the way down to Platinum and Goodnews Bay on the sand. And they also grow on the sand extending up the coast.

Martha Mark: But they don't grow at Uiluq [River] because it's mud.

Pauline Matthew: Those seem to grow in June, toward the end [of June]. And when they get hard, they stop gathering them. Then they grow a second time. So when they get hard or when those beach greens start to grow flowers [they stop picking them].

George Pleasant: When they start to light up, *kenurrangaqata*.

Martha Mark: Yes, they say they start to light up, *kenurrangluteng*, when they start to turn yellow.

Pauline Mathews: When they start to turn a little yellow, they refer to them as *kenurrangluteng* [starting to light up].

Marsh marigolds[8]

Martha Mark: Then along the shores of sloughs are *allngiguat* [marsh marigolds].

Pauline Matthew: Yes, I used to obtain them from up there. Along the shores of streams, those marsh marigolds grow along the shores of streams and have small yellow things on them. They used to gather those marsh marigolds before they bloomed. They ate them when they cooked half-dried and boiled fish.

It'garalget

Tartuilnguq: Tua-ll' kapuukaraat pelluata it'garalegnek. It'garalegnek avungluta ayalallemni.

It'garalget-llu call' aipiriluteng nautuluteng. Ukut ciuqliit enrirluteng tegg'iluteng *yellow*-aanek-llu pingluteng. Tua-i-ll' July-ami allat nauluteng avayaitgun. Assiraqluteng unaunateng.

Miisaq: It'garalget unani cenami *all the way down to Platinum and Goodnews*, cenani tamakut tamarmeng qaugyam qaingani nautuut. Qavatmun-llu tua-i cali qaugyam qaingani nautuluteng.

Tartuilnguq: Uilumi taugaam piyuunateng marayaungami.

Miisaq: *June*-ami tamakut pingatelartut *toward the end*. Tuall' enriata pinrirluki. Aipiriluteng cali naunqiggluteng. *So* enringa'artaqata wall' naucetaanek nauginga'artaqata tamakut it'garalget.

Arnariaq: Kenurrangaqata.

Tartuilnguq: Yaa kenurrangnilarait *yellow*-aangaqata.

Miisaq: *Yellow*-aacuaranga'artaqata kenurrangnilarait.

Allngiguat

Tartuilnguq: Tua-i-ll' augkut kuigaaraat cenaitni allngiguat.

Miisaq: Yaa, pit'lallruunga pagaaken. Kuiggacuum ceniini tamakut, kuiggacuum, *like stream* ceniini tamakut nautuut allngiguat *yellow*-rrarnek waten pingqerrluteng. Tamakut-wa naucetaarurpailgata pitetullruit, *bloom*-arpailgata allngiguat cali. Egamaarrlugnun atutuluki.

Martha Mark: They are good to eat with half-dried and boiled salmon.

Pauline Matthew: Along the shores of a river, along their streams. But these days, many people no longer gather those.

Fiddlehead ferns

Martha Mark: *Ceturqaaraat* [fiddlehead ferns, wood ferns] also grow outside. When they do this, when they start to open up, they are good to gather.

Pauline Matthew: We gather those in May, toward the end of the month. Those are made into *akutaq*. The only way that I use those is to make them into *akutaq*.

Martha Mark: The only thing I do is make them into *akutaq* also.

Pauline Matthew: But it's good when picking them when they are just about to open up. If you pick them before they open up, it's really tiring to clean them. But I obtain them from these places; they are usually along places that are caved in. You know there's land, then here where it suddenly goes down, around those places.

George Pleasant: Yes, along the edge of where the tundra suddenly descends and becomes marshland.

Pauline Matthew: Along a place that suddenly drops, there are many along those places. I gather some from behind the old airport. And there are a great many at Ingricuar.

Martha Mark: And at Platinum, from across there.

George Pleasant: There are also many of those down below the new airport, beyond the edge where you go up.

Pauline Matthew: I also obtain those from down the coast. You travel along Agalik River, it heads down; those places that have a lot of bushes, they are mainly abundant around those places. I mainly gather those from around those places. They are obvious, and there are large numbers mostly in those areas.

Tartuilnguq: Egamaarrlugnun assirluteng.

Miisaq: Kuigem ceniini, iciw' *stream*-aitni maani. Taugaam tamakut maa-i amlleq pissiyaayuirutait.

Ceturqaaraat

Tartuilnguq: Ceturqaaraat-llu qagaa-i qagaani pituut. Waten imumek piaqata, pet'ngartaqameng pic'unaqluteng.

Miisaq: May-ami tamakut picetuaput, *toward the end of the month*. Tamakut-llu tua-i akutautuluteng. Akutauluki wiinga atulaqenka kiingan.

Tartuilnguq: Wiinga-ll' kiingan akutelaqenka.

Miisaq: Taugaam assirluteng imumek tua-i pet'ngeqatarluki *pick*-allrit. Petengvailgata *pick*-aquvet carrilngunaqpiartut. Taugaam makuni pit'lartua wii; merignerni atam uitatuut. Iciw' waten nuna man'a tua-ll' waten atraqertellrani, tamaani atam.

Arnariaq: Yaa, inigutmi.

Miisaq: Atraqallrani, tamaani atam pilituuq. Pikaken-wa *old airport*-am keluanek, tamaaken pit'lartua. Ingricuarmi-ll' tua-i nutaan amlleqapiat.

Tartuilnguq: Platinum-aami-llu agaaken.

Arnariaq: *New airport*-am-wa cali ketiini yaani mayuryarami aug'um mengliin yaatiini cali amllertut tamakucit.

Miisaq: Un'gaaken-llu cali tamakunek picetuunga. Iciw' Agaligkun ayagluni, ayalria iciw' un'gaatmun ayalria; tamakuni cali augkut, napalquyagalilriit iciw' augkut, tamakut nuniitni amllerpallulartut. Tamakut nuniitni piterpallulartua tamakunek. Nallunaitelartut amllerrluteng tamaani *mostly*.

Martha Mark: The roots of *ceturqaaraat* are *kun'at* [edible roots of spreading wood fern]. The small things are long. Those taste very sugary. You dig underneath [the fern], and then underneath they are very small and white. *Kun'at*, the roots of *ceturqaaraat*; its root.

George Pleasant: They are good cooked.

Martha Mark: They taste very sugary and are delicious.

George Pleasant: I compare their taste to potatoes. When digging them out before they grow, those are good when cooking them. They dig and obtain them before the plant above them grows. They cook them, cooking their roots.

Martha Mark: It's good to dig them and get them, even when they're grown.

Sea lovage

Martha Mark: And those *mecurtulgaat* [sea lovage]; *mecurtulgaat* grow around *taperrnat* [coarse seashore grass]. Along the ocean shore.

Pauline Matthews: They only eat them when eating half-dried and boiled fish.

Martha Mark: Yes, they are good. And they are good made into *akutaq*, adding raisins to them. And a certain part of them, they are very good when just eating them fresh without cooking them.

You know the branches of *mecurtulgaat*, and then there are these [leaves]. They eat them. You know how they grow like this, and then we take their branches and we take those [leaves], too, and eat them.

When they had just grown; [later] they get hard and become inedible.

George Pleasant: *Mecurtulit.* When I see some down along the ocean shore, I cut them and eat them. Just eating the ones that have just grown without cooking them is also good.

Tartuilnguq: Ceturqaaraat aciit kunaugut. Takyaarluteng. Saarralarninaqpiarluteng tamakut. Lagluku acia, tua-i-ll' waten acia qatyaarluteng waten. Kun'at, ceturqaaraat aciit im' waten; acia.

Arnariaq: Kenirluteng assirtut.

Tartuilnguq: Saarralarninaqpiat aspiat.

Arnariaq: Wiinga-am *potatoes*-aacetun ayuqekutelaranka. Nauvailgata elagluki assirtut tamakut kenirluki. Augkut pikani naulriit nauvailgata elagluki pitaqluki. Kenirluki, kenirluki *root*-ait.

Tartuilnguq: Naullrungraata assirluni elagluki pill'uki.

Mecurtulgaat wall' mecurtulit

Tartuilnguq: Augkut-llu mecurtulgaat; taperrnat nuniitni nautuut mecurtulgaat. Imarpiim ceniini.

Miisaq: Egamaarrlugnun kiingan aturluki.

Tartuilnguq: Ii-i, assirluteng. Akulluki-ll' assirluteng, *raisins*-iirluki. Imkut-llu waten piit, qangqurluki aspiat kenirpek'naki.

Iciw' makut avayait mecurtulgaat makut-wa. Nerluki. Iciw' tua-i waten naulalriit tua-ll' tamakut avayait teguluki tamakut-llu piit call' teguluki ner'aqluki.

Naugaaruaqata; enrituut, tegg'ituut, pisciigaliluteng.

Arnariaq: Mecurtulit. Tamakut wiinga-ll' unani cenami tangrraqama kepurluki nerlaranka. Keninrilngermeng nerluki cali assirtut naunerraraat.

Tall cottongrass[9]

Alice Rearden: What about those *iitaat* [tall cottongrass plants]?

Theresa Abraham: There are many *iitaat* around.

We eat them, and we also eat [their tubers] when voles gather them in their underground caches.

They gather those *iitaat* down on the coast, then they remove the edible part, and then they soak the edible section in water; and then they eat them by dipping them in seal oil. Then they used [the grass section] for braiding [fish], or they also used them for weaving and braiding in the past.

Mare's tail plants[10]

Pauline Jimmie: The spring season was of utmost importance to our parents. Then during the fall, they'd gather coarse beach grass, they'd scatter some coarse beach grass and coarse grass. When it got cold out, we would go and pry the ground to look for mouse food, for grass seeds and tubers of tall cottongrass. Then at freeze-up, we'd collect mare's tail plants.

My father told me that I should always keep a supply of mare's tail plants. He said that I should always gather a lot of mare's tail during the fall, even if it was a large amount. He said mare's tail mixed with seal oil is a food that can be eaten during starvation times. He said long ago when they experienced famine, mare's tail plants were eaten. My grandmother fed those people mare's tail plants, but mainly mixed with seal oil.

He told me that I should gather some mare's tail plants, even if it was a large amount, that it was okay if I discarded the leftovers during spring when birds were around. That's why I always try to gather mare's tail plants during the fall, right after freeze-up. That person and I, with each other's help, always try to fill two plastic bags, even if it's a large amount. I discard them during summer when I know that we won't go hungry.

Iitaat

Cucuaq: Augkut-mi iitaat?

Paniliar: Iitaat amllertut.

Nerluki, uugnaraat tuaten neqliaqatki ner'aqluki.

Tamakut iitaat unani pitaqluki tua-i-llu iiciluki, tamakut-llu iitait miiqerluki uitalluki; tua-i-ll' nerluki uqiqaqluki nerluki. Tamakut-llu augkut piirritekluki wall'u tupigluki pilallruyaaqekait, piirritkaqluki.

Tayarut

Kangrilnguq: Up'nerkaq-wa un' arcaqerluku pitullrukiit augkut angayuqaamta. Tuamtell' uksuarmi taperrnarnek, taperrnarnek cagciqunaurtut, kelugkarnek. Tua-i-am nenglengareskan pakissaagluta uugnaraat neqaitnek, utngungssarnek, anllernek. Tuamtell' cikuqaqan tayarunek, tayarulluta.

Wiinga-ll' im' aatama pillrukiinga tayaruutaiteksaunii pikilii. Uksuaraqan-gguq tayaruquraraqlua amlleringraata. Tayaruq-gguq man'a kaigem nalliini neqnguuq uqumek avuluni. Ak'a-gguq tamaani kaillratni tayaruq man'a neqngullruuq. Taum *grandma*-llma tamakut tayarunek neqillrui uqurpalluuluki taugaam uqumek avuluki.

Tayarutaqlua pilaasqellua amlleringraata, up'nerkarqan waten yaqulget piaqata egtengramki canritniluki. Taumek waniw' tayarutengnaquralalrianga waten ukusarmi cikuqerqan. Ikayuqlunuk ikna-ll' amlleringraata *plastic*-aak malruk imirlukek pitengnaquraraqlua. Kiagan tua-i egtaqluki kaigngairutaqamta.

Wild rhubarb[11]

Peter Jacobs: What about when the wind is always calm during winter? I think there is a saying about that. You know the wild rhubarb would get as tall as a person, the sour dock's mate, because there would be no wind.

The frost, I think that has an explanation. You know when the wind has been calm for so long, and soon the blackfish traps would be filled only with *mineq* [?algae] inside because it had been windless. Is that saying true? Please talk about it.

Frank Andrew: They say when the world was in its original state and was cared for and respected, and not contaminated, that was how the world was. And all fish were abundant. And the month of January, the month after it that they call Kanruyauciq on the coast, the month of February, they said it was a month what *?pinaruaryaraqluku* since the beginning [of time].

They say that in January, when the weather was in its original state, the wind would stop blowing as usual. And the wild rhubarbs would get as tall as people because of the frost, because there was so much frost. That's why our shorefast ice on the ocean used to be very long; it would be far out there and would be down near the edge of deep water. That's why hunters apparently were successful catching sea mammals when it was like that. They said it was like the people of the coast were at spring camp hunting seals when it would *?pinaruaraqan* in February.

Peter Jacobs: When it wasn't windy for a period of time?

Frank Andrew: Yes, they say it used to get warm in February, and there would be a lot of water, and the lakes would fill with water. And even though it got cold in March, before it got too cold, it began to turn into summer.

Mosses and mosses aged in seal oil[12]

Lizzie Chimiugak: I have now become an old woman before I even made *puyat*. Those *puyat* [mosses aged in seal oil]. Long ago our dear ancestors regarded them as essential commodities. And I saw them, they used to have a supply of *puyat*. They were never without *puyat*. They put *puyat* in seal stomachs and let them soak in seal oil, and I don't know how long they let

Nakaaret

Paniguaq: Una-mi uksumi quunirturassiyaagaqami? Cal'
qanertangqerrsugnarquq. Iciw' imkut-llu nakaaret yugtun imkut quagcit
piit yugtun angtariluteng ugaani anuqliyuitem.

Kanrem, tamana-ll' qanertangqerrsugnarquq. Iciw' anuqliyuunani tua-i
kiituani can'giircuutet-llu camkut minermek-ll' tam' imangyaurrluteng
quunirturassiyaagaqan. Tamana-qaa piciuguq? Kitek' qanemcikqerru.

Miisaq: Ella-gguq ellaullrani kencikumatuluni uqlarpek'nani, imna
atullratni, tuaten ayuqellruuq ella. Neqet-llu tamiin tua-i amllerrluteng.
Tauna-llu tua-i una *January*, Kanruyaucimek un'gani aterpagtaatukiit,
tunglia una *February*, nutem-gguq pinaruaryaraqluku tua-i.

Tuani-gguq tua January-mi ellaullermini anuqlikegtayuirutetuuq.
Nakaaret-llu yugtun angtariaqluteng kanrem, kanrem ugaan'. Taumek
un'a imarpigput tuar acetutullrullinilria yaaqsigluni camani tua-i, iginam
nuniini camani uitaluni. Taumek elluatun unguvalriartelallrullinilriit
tua tamakut pingnatulriit tuaten ayuqellrani. Cenami un'gaani tua-i
cenamelnguut-gguq pinaruaraaqan February, upnerkilriatun tua-i
ayuqlirituut unguvalrianek tua-i imkut.

Paniguaq: Anuqliyuitaqan-qaa?

Miisaq: Ii-i, nenglairulluni-gguq tua-i pituuq *February*-mi, mel'ingluni-llu
miiqangluni nanvat miiqangluki. Tua-i-llu nenglengyaaqeng'ermi *March*-
aami maani pacepailgatki kiagyungluni.

Urut puyat-llu

Neng'uryar: Puyiqatainanemni arnassagartua wiinga. Im', puyat.
Tamaa-i ak'a ciuliaurlumta-ll' cacetuqutkacagallrit. Wiinga-ll' tangssugluki
puyautengqerraqluteng. Puyautaicuunateng. Anrutanek caquluki puyat
uqumek mecirluki qaillun-llu tayima mecilallruitki. Anrutanek *plastic*-at
mikuitellratni tangerqerraalallrulrianga puyautnek.

them soak. I first saw *puyat* in seal stomach containers when plastics were not readily available.

Nuuniq [Louise Ayaginar] taught me how, taking a *puyaq* like this. How did they eat these when they didn't make them into *akutaq*? When she did [eat some], she plucked some down feathers from my parka, and she told me to mix it [with *puyat*] and chew on it, and she also chewed. She said they ate *puyat* like that long ago when they'd go through starvation. Nowadays, they have five-gallon plastic containers.

Sophie Agimuk: They used to make *puyat* using those *urut* [mosses] they had collected.

Lizzie Chimiugak: I have planned on making *puyaq* on my own, and I have thought about it.

John Alirkar: They would make *akutaq* with them—some of them "fly away." [*laughter*]

Sophie Agimuk: After they picked [the mosses], they dried them. And then when they dried, they soaked them with new seal oil.

Phillip Moses: [*chuckling, as he is remembering a hilarious incident that happened to his sister*][13]

Sophie Agimuk: And after they soaked them, they put them in their spot and left them there. And then those pieces of moss got so saturated with jelled oil that they became thick and sticky. That was the way they made [*puyat*].

Phillip Moses: [*chuckles again, look out, it's coming*]

Sophie Agimuk: They collected these pieces of moss and dried them, and when they had thoroughly dried, they poured seal oil over them and left them like that.

Phillip Moses: [*chuckles, the volcano coming to a boil*]

Sophie Agimuk: When they leave them alone like that, they get sticky and they are extremely moldy, and thus it is a very sticky substance. When it becomes like that, it becomes edible.

Nuunim call' alerquallrukiinga waten puyaq teguluku. Qaill' makut nerlartatki akutevkenaki? Piami atkuunka waten piqerluki yaqulget qivyuitnek vegtaaraqerluni avuluki tamuasqelluki ellii-ll' tamualuni. Waten-gguq ak'a nutem nerlallruit-llu paluyugaqameng puyat. Maa-i assigtaqegciyaaqelriit *plastic*-anek *five gallon*-anek.

Avegyaq: Puyitullrulriit imkunek urunek pilluteng.

Neng'uryar: Puyiarkautuyaaqua-am wangnek-llu umyuarteqlua.

Allirkar: Akutaqluteng—ilait teng'aqluteng. [*ngel'artut*]

Avegyaq: Piq'aaluteng-llu kinercirluki. Kinqaki-llu nutaan uqumek nutarinrarmek mecirluki.

Nurataaq: [*ngelaq'erluni*]

Avegyaq: Mecirraarluki-llu tua-i nekaatnun tua-i elliqerluki uitauraraqluteng. Kiituan tua-i taukut wani urut imuurtut tua-i puyak'acagarluk' tua-i nepcanariluteng. Tuaten pilitullruut.

Nurataaq: [*ngelaq'erluni*]

Avegyaq: Urunek makunek cumerrluteng kinercirluki, kinenqegcaarqaki uqumek kuvluki uitalluki.

Nurataaq: [*ngelaq'erluni*]

Avegyaq: Tua-i uitaskaceteng tuaten nepcanariluteng puyak'acagarluki tua-i im'uluni tua-i nepcanaqluni. Nutaan tuaten pikan nutaan neryugngariluku.

John Alirkar: And they used them on kayaks that formed open seams to make them more waterproof.

Sophie Agimuk: It becomes extremely sticky.

Lizzie Chimiugak: It was because those poor people used everything; one would also apply *puyaq* on one's kayak.

Sophie Agimuk: They used to do that. Instead of using old seal oil, they used new [oil].

David Chanar: When do they eat it?

Phillip Moses: They made *akutaq* out of it. [*laughter, it's getting close*]

Sophie Agimuk: And when they mixed them, they kept mashing it by adding water to it. Soon that mixture would multiply. When it became enough, they mixed it with blackberries.

So they become *akutat*, and they become edible.

But as soon as one finishes [making *akutaq*] out of it, it must be eaten right away. And if the *akutaq* is left alone for a short while, it becomes watery and the berries inside float on top, it would become all water.

John Alirkar: Just like ice cream. So they say it "takes wings."

Sophie Agimuk: It became all water.

Phillip Moses: It takes wings.

David Chanar: It flies away?

Lizzie Chimiugak: They said that it would fly away, and not to leave any of it uneaten—they said that it would fly away. And believing them, I thought that it would be floating up there [in the sky].

Phillip Moses: I watched my older sister, Arnaqulluk, down there at Qalulleq or rather up there when she made *akutaq*, and it was apparently that kind of *akutaq*. This was long before she got a husband, and she must

Allirkar: Qayanun tuaten aivkalrianun umcigutekluki.

Avegyaq: Imuuluni tua-i nepcukaraurrluni.

Neng'uryar: Ca tamalkuan ateurlullruamegteggu; qayani-ll' kia puyataqluku.

Avegyaq: Tuaten-am pitullruut. Ak'allamek-am uqumek pivkenaku taugaam nutaramek.

Cingurruk: Qaill' piaqameng ner'aqluku?

Nurataaq: Akulluku. [*ngel'artut*]

Avegyaq: Akuskunegteki-llu tamakut passiurluki merr'armek kuvqaqluki. Kiituan' tauna amlleriuq. Amllerikan-llu tan'gerpagnek.

Akutaurrluteng tua-i nernariluteng.

Taugaam tua-i uitatevkenaki egmian' taquciicetun ner'arkauluki. Tua-i-ll' akutaq tauna uitacuaqaquni imuuluni tua-i mer'urrluni qiviminek pugtaqerluni, merrlainaurrluni.

Allirkar: *Ice cream*-atun. Tengluni-gguq tua-i.

Avegyaq: Merrlainaurrluni.

Nurataaq: Tengluni.

Cingurruk: Tengluni-qaa?

Neng'uryar: Tengciqniluku ilakuisqevkenata pitullrukiikut—tengciquq-gguq. Wiinga-llu ukverlua pagaanetnayukluku.

Nurataaq: Tangssullruunga wii alqamnek, Arnaqullugmek, uani ua-i, Qalullermi kiani akutellria cunaw' tamatumek akutellinilria. Uingyugnaunani, nasaurluuluni pillilria tayima taugaam angturriluni.

have been a young woman then, but she was grown up. So she was making that kind of *akutaq*. Pitching a tent, we used to stay there every once in a while.

So it was at about that time. She was making *akutaq*.

So when she added berries to it, she found that they were not enough, there weren't enough. So covering her *akutaq*, the *akutaq* she was making, she hurried up above Qalulleq to get more berries since there were usually blackberries there. So she hurriedly went to get some to add [to her *akutaq*].

So eventually she came in. So when she came in, she [went to] her *akutaq* [and uncovered it]. [*laughs*]

It had become nothing but water!

It had flown, it had flown as they say! Probably because she was seeing that happen for the first time, sitting down, she removed the cover and, suspecting her sisters and also the father of Cacungaq, [she exclaimed], "Now who poured water into this *akutaq* I was making?" [*laughter*]

"Now who in tarnation poured water into my *akutaq*!" [*laughter*]

So my mother told her, "You see, since you left it to get berries to mix into it, it had flown, as they say." [*laughter*]

When I think of that sometimes, when she did that in my presence [asking] who had poured water into it. [*laughter*]

Lizzie Chimiugak: And they talk about that certain individual who, when told that it had flown, he looked up, asked where it was, and even waved his arms around in the air.

Phillip Moses: So those people were like that, and now these two have related how they start making *puyaq* by pouring water a bit at a time, that is the way they make those kinds of *akutat*. So they would periodically pour a little bit of water into it. So when it was time, they mixed blackberries into it. So they eat them right away, and they are delicious. So they were like that.

Akulluni. Pelatekarluta tuani uitatullruamta caaqamta.

Taum nalliini. Akulluni.

Tua-i-ll' qiviryaaqekni qivii ikgetqerluteng, ikgetqalliniluteng, ikgetqerluteng. Patuqerluku tua-i akutani, akuskengani tauna patuqerluku qivirkarrsulaalliniluni pikaken pagna tan'gerpangqelaan Qalulleq. Aqvat'laagluni qivirkaminek.

Tua-i iterluni. Itrami tua-i tauna akutani taukut [patuirluki]. [ngel'arluni]

Mer'urtellinilria!

Tengllilniluni, apqiitnek tengllilniluni! Nutaan tua-i tangssulliami taumek aqumngami patuiraa ukut alqani Cacungaam-ll' atii kacikluki, "Kia tanem' una akutaq kuvvliniagu mermek!" [ngel'artut]

"Kia tanem una akutaqa mermek kuvvliniagu!" [ngel'artut]

Aanama pia, "Tua-i-gguq unilluku qivirkarrsuavgu teng'uq." [ngel'artut]

Tauna caaqama umyuaqaqamku takuk'acagamni kitumun mermek kuvellranek. [ngel'artut]

Neng'uryar: Kina-llu-ggem tauna tengnillrani ciuggluni nauggaaraluni qilaliurlun' tuaten.

Nurataaq: Tuaten taugaam tua-i ayuqetullruut, augkut-llu ukuk qanemcikek puyamek ayagnirluteng mermek kuvqaqluku, akuquratullruut. Tua-i mermek amllenrilngurmek kuvqaqluku. Pinarikan tua-i tayima tangerrluku nutaan tan'gerpagnek qivirluku. Egmiin nerluteng assirluteng neqniqluteng. Tuaten-am tua-i ayuqellruut.

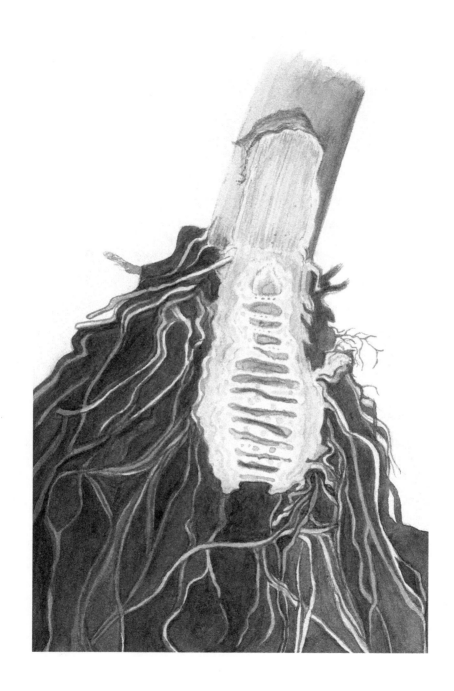

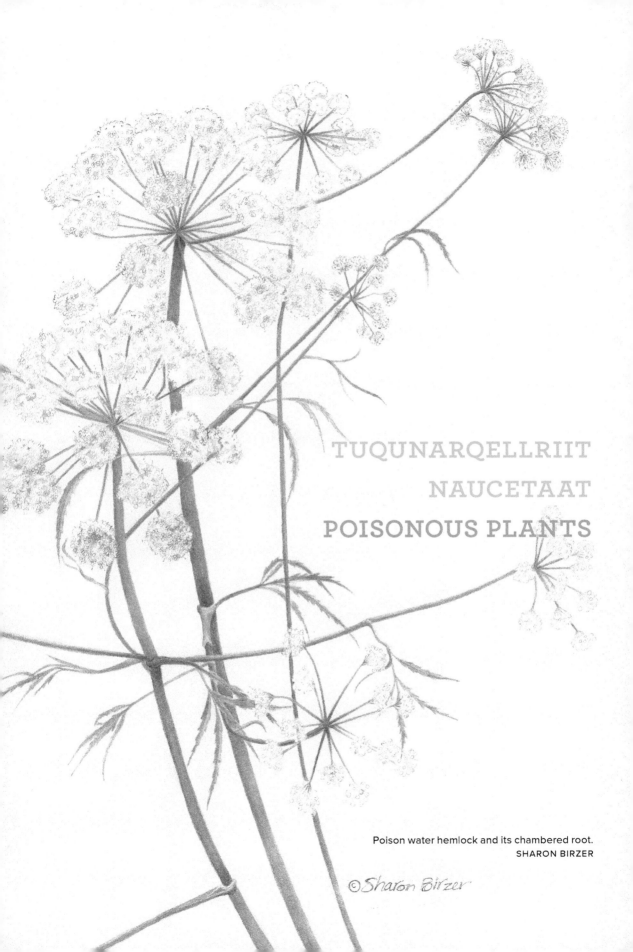

TUQUNARQELLRIIT

NAUCETAAT

POISONOUS PLANTS

Poison water hemlock and its chambered root.
SHARON BIRZER

Male water hemlock[14]

Lizzie Chimiugak: On top of the mountain, there was a very large den up there that they had never gone to see. With a mountain behind them, there was a married couple living there, and they were alone.

Then one day, her husband told his wife to prepare a light for him. "What are you going to do [she asked]?" "Let me go and look at the inside of that den up there."

Then she had him bring a light, he went inside there with a light. He went inside with a light, and [he saw] the very first thing.

As he was walking, he came upon a shrew, a very small one.

Since this story is a *quliraq* [traditional tale in which amazing things happen], he asked [the shrew], "Where is the scary one?" It replied to him, "*Qama kangia* [In there at the source]." They were getting larger and larger, and the shrew also, the ones he was meeting got larger and larger as he went toward his destination's end. He'd ask them, "Where is the scary one?" They would all reply, "*Qama kangia*." They would say that it was at the top.

The animals that he came across got larger and larger, all the ones that he came across got larger, and he constantly asked them and they continually replied with *qama kangia* [in there at the source].

Then they say he came upon a door.

He continually came across all the fur animals, and they got larger and larger. I wonder what he came across last, maybe a *quugaarpak* [mammoth].

He came upon all the animals as he traveled for a long time while he was walking along holding a light, and his light never went out. He came upon many animals, and when he'd ask them, they always said *qama kangia* to him.

After he had been walking for a long time . . . that married couple was young. He came upon a door. When it appeared, there was someone lying on his back along the back area with his stomach slashed open, a person. They say his limbs were scary.

Anguteurluq

Neng'uryar: Ingrim, ingrim qacarnerani igterpakayall'er ping' paqcimayuunani. Ingrimek kelulutek, taukuk tua-i nulirqelriik uitallinilriik kiimek-llu.

Tua-i-ll' erenrem iliini angutii kenurrarkiuresqelluni aiparminun. "Caqatarcit?" "Pingna, pi wani, igtem ilua paqteqernaurqa."

Tua-i-ll' kenurramek tua-i ayaucelluku, kenurrarluni taun' iterluku. Kenurrarturluni itliniluni, ciukacagarmek.

Pairkengyartuqili angyayagarmek, imumek mikcuayaarmek.

Tua-ll'-am qulirnaungami aptelliniluku, "Naam alingnarqelria?" Kiulliniluku, "Qama kangia." Angliriinarluteng uugnar-llu, angliriinarluteng iqutmun paireskengai. Aptaqateng, "Naam alingnarqelria?" Waten kiugurluteng, "Qama kangia." Kangiani uitaniluku.

Tua-i ungungssit angliriinarluteng pairtellni tamalkuan angliriinarluteng pairqurallra, apqurluku qama kingiarturluku.

Tua-i-llu-gguq amigmek tekilluni.

Tamakut tamaa-i nunam melqulgi tamaita pairrluki, angliriinarluteng. Camek-kiq nangnermi pairkengellrua, quugaarpagmek-llu pilliuq.

Ungungssit qaqilluki paircimakai-gguq kenurrarluni ayallermini, kenurraa-ll' nipsuunani. Ungungssirugaat pairrluki, aptaqateng qama kangiarturluku.

Ayalnguami tua-i . . . ayagyuallrulutek tua-i taukuk nulirqelriik. Amigmun tekicartulliniluni. Pugluni piuq qaugna taklalria egkuani aqsiig ullingqalutek, yuk-wa tua-i. Tua-i-gguq cangilluteng makut avayai.

Then they say when he ran out, he dropped his light and ran, he ran when he suddenly became frightened. Not long after, he reached the opening [of the den]. When he reached the opening, there was no house along the place where their house had been located.

When he went down, their old elevated cache was completely rotten, and their old home was there, and his dear wife was gone. And its old posts were extremely rotten.

Then since he didn't understand what was going on, since he suddenly became thirsty, he went up, and when he was about to take a drink from a small lake, he put his head down; he saw that his reflection was that of a very old man. While he was walking along there, he had become an elderly man. But when he ran home, he had quickly reached [the den's] opening.

Ruth Jimmie: Maybe those [animals] represented years.

Lizzie Chimiugak: My, I wonder if the ones [fur animals] that he came across were years.

Albertina Dull: My, that's probably so, it's probably a year.

Lizzie Chimiugak: He came across many, many things. When he'd ask them, they would continually say *qama kangia* to him.

He was evidently an elderly man, his reflection down there was elderly.

Since he didn't know what to do, he suddenly fell on his buttocks in the water along the shores of a lake, and was an *anguteurluq* [male water hemlock plant, lit., "poor dear man"], he became an *anguteurluq*. Those *anguteurluut* that grow that have poisonous tubers; as you know lakes have them.

He suddenly sat down there. "Let me be one of these forever, so that in the far future, they will harvest me." He became one of those, since he didn't see his wife; the poor thing probably died, she probably starved to death.

While he went through [that den], he evidently became a very old man.

Tua-i-llu-gguq anqertellermini kenurrani peggluku aqvaqurluni, aqvaqurluni, alingqallermini. Ak'anivkenani-gguq pailliarrluni. Pailliarcaaquq-gguq tang kan'a enellrak enetaunani.

Maaten-gguq atrarluni pilliniuq mayurrvillrak-llu un' aruk'acagaumaluni, enellrak-gguq-wa una, nauwa-w' aipaurlua. Naparyacillri-ll' aruk'acagarluteng.

Tua-i-llu-gguq uunguciilami tagluni meqsulliimi nanvarra'armun meq'atarluni put'uq; tarenraa kan' angulluakacagarluni. Tamaaggun ayallermini angulluarurtellrullliniluni. Taugken-gguq aqvaqurluni utertellermini tamaa painga tekiarrluku.

Angalgaq: Tamaa-i-w' tamakut allrakuullrullilriit.

Neng'uryar: Ala-i allrakuullrullius-kiq tamakut pairqurallri.

Cingyukan: Ala-i pillilria-wa, allrakuullrullilria.

Neng'uryar: Pairtellri amelkacagarluteng. Aptaqateng qama kangiarturluteng.

Angulluallinilria, angulluarluni kan'a tarenraa.

Tua-i cacirkailami maavet nanvam ceniinun aqumkallalliniluni mermun, anguteurluuluni, anguteurluurrluni. Imkut naumatulriit anguteurluut aciit *poison*-at; nanvat pitangqetulriit.

Tamaavet aqumqertellliniluni, "Makuugurlii, ak'aku kinguliaku pitaqelarniaraatnga." Tua-i-gguq tamakuurrluni, nuliani-ll' tangenrrilamiu; tuqu'urlullrullilria, palullrullilria.

Angulkacagallinill' tamaaggun ayallermini.

NAUCETAAT
IINRUKTUKNGAIT
MEDICINAL PLANTS

The entire surface of the land is medicine[15]

Barbara Joe: They would gather plants in the summer. Plants were their foods. In the spring, we'd start eating *iitat* [lower parts of tall cottongrass], our mother would go and gather *iitat*. Then when these greens grew, again . . .

Ones that they used as medicine. And these *quagcit* [sour dock], they say sour dock also, they cut them into pieces, and they are extremely tasty made into *akutaq*.

They are very delicious. Those plants prevented us from being ill.

Also, when they would open up salmonberries in winter, when we'd have colds, they would have us drink berry juice in the morning. And not filling the cup full, they would let us drink a small amount of berry juice. They say those are medicine. Since those things were medicine to their bodies, since they have many vitamins, they say those ones, today I've come to understand what they're like, they say eating those berries makes one strong since they have many vitamins.

Today, we poor things today, since we no longer eat those things we were raised on, we tend to get weak because we cannot go and get them ourselves.

They would let us eat *iitat*. And when plants grew, they would eat plants until they became hard. And when salmonberries were ready as well as sour dock, they also had us eat sour dock. And when they had picked a lot of salmonberries, they would fill a barrel with salmonberries. But the leaves of sour dock, they placed the sour dock leaves opposite each other and covered them [with berries]. And [the berries] never became rancid.

Those things they raised us on, some of us happened to catch those things. They say their foods prevented them from becoming ill all the time. They continually ate their foods and became elderly.

And *kapuukaraat* [buttercups] also. And those *arnaurluut* [wild rhubarb]. We used to eat *arnaurluut*. In lakes, their leaves look like *caiggluk* [wormwood] leaves in lakes. We would gather those.

Peter Black: *Nasqupaguat* [nutty saw-wort] also?

Yungcautnguuq nunam qainga tamarmi

Arnaucuaq: Naunranek kiagmi cumerrluteng. Naunraat taugaam neqekluki. Up'nerkami iitarnek, iitarnek nerengnaurtukut, aanavut iitarcurnaurtuq. Tua-llu makut *greens*-at ataam naungata . . .

Ellaita yungcautekellrit. Quagcit-llu makut, cali quagcit-gguq tua-i, itumqurluki akutauluteng assiqapigtut.

Assiqapiggluteng. Tamakut naunraat wangkuta nangteqsuitevkallruitkut.

Cali atsat uksumi angpartaqamegteki qusraqamta atsat *juice*-aitnek unuakumi mercetnauraitkut. Caaskaq-llu muirpek'naku meq'ercetnauraitkut atsat egenritnek. Tamaa-i-gguq tamakut yungcautnguut. Tamakut yungcautekngamegteki qaimeggnun, *vitamin*-aliata, tamakut-gguq, maa-i nutaan taringlaranka, naunraat-gguq tamakut neqkellrat kayunarquq *vitamin*-alirlaata.

Maa-i wangkurlumta maa-i tamakut neqkenriamteki kayuirucugtukut anglicautelput imutun wangkutnek aqvauqurciigaliamta.

Iitarnek tua-i nerevkarnauraitkut. Naunraat-llu pic'ata tua-i naunraat enrillratnun cali neqekluki. Atsanek-llu ataam piurcata quagcit-llu, quagcinek cali neqengqercelluta. Tua-i-ll' atsauciameng, napartamun atsanek, napartaq imirnauraat atsanek. Tamakut taugken quagcit cuyaitnek kipullguqu'urluki patuluki. Uquggluyuunateng-llu.

Anglicautelteng, anguqallruaput ilaitni maani tamakut. Tamakut-gguq neqaita nangteqvakarcecuitellruit. Tua-i neqet anglicautelteng tamakut neqkurluki teggenrurtetuluteng.

Kapuukaraat-llu. Imkut arnaurluut-llu. Wangkuta arnaurlurtutullruukut. Nanvani cuyait tuar imkut caiggluut, caiggluut tua-i cuyait nanvami. Tamakunek pissurnaurtukut.

Nanirqun: Nasqupaguat-llu-qaa?

Barbara Joe: Yes, nutty saw-wort.

Maryann Andrews: *Nasqupaguat* and the others like those *nasqupaguat*, what do they call them again, *mecuqelugat* [sea lovage]?

Barbara Joe: Yes, those were our foods. They used them to heal themselves; since those were their foods, they continued to have us eat them.

Salmonberry juice, and when crowberries would get juice, as you know sometimes when they pop, they get juice; they didn't discard those either. They would have us eat them, they had us eat them for medicine.

Peter Black: And the tubers of *iitat* also?

Barbara Joe: Yes.

Maryann Andrews: And the tubers of *iitat* are delicious with seal oil. . . .

Those *uqvigat* [willows], what do they call them, and they use them as medicine?

Peter Black: The bark of *uqvigpiit* [Alaska bog willows].

Alice Rearden: I'm asking about bog willows, whether they used them as medicine or ate them.

Maryann Andrews: Their juice.

Barbara Joe: What kinds of willows? We call those that have markings on them *cuukvaguat* [alders]. And the other kind, we call them *cuyakegglit* [diamond leaf willows], the ones with orange tops. And then there are also *uqvigpiit* [Alaska bog willows]. Which one are you talking about?

Alice Rearden: I don't know those, how did they use them as medicine, did they use them all?

Arnaucuaq: Ii-i, nasqupaguat.

Tauyaaq: Nasqupaguat imkut-llu aipait cali nasqupaguat, caneg'-ima piqerlaqait, mecuqelugat?

Arnaucuaq: Yaa, tamakunek neqengqerrluta. Ellaita yungcautekluki; neqeklallruamegteki egmiulluki neqketullruit, wangkutnun nerevkatullruit.

Atsat tua-i egenrit tan'gerpiit-llu egnengaqata iliikun qagertaqameng egnenglalriit; tamakut-llu egcuunaki. Wangkutnun tua-i nerevkaraqluki yungcautekevkarluki.

Nanirqun: Iitat-llu imkut aciit camkut?

Arnaucuaq: Ee-m.

Tauyaaq: Iitat-llu aciit neqniqluteng uqirluki. . . .

Imkut-qaa uqvigaat camek-im' piqerlaqait, yungcautekluki-llu?

Nanirqun: Uqvigpiit amiit.

Cucuaq: Uqviganek aptua qaillun iinrukluki wall'u-q' nerluki pitullritnek.

Tauyaaq: *Juice*-ait.

Arnaucuaq: Cakucit uqvigaat? Wangkuta cuukvaguanek imkut pilaraput qaralingqerrluteng. Aipaingit-llu cuyakegglinek, kangrit *orange*-auluteng. Ataam-llu makut uqvigpiit. Naliitnek qanercit?

Cucuaq: Tua-i-w', nallukenka tamakut, qaillun iinrukluki, tamarmeng-qaa pitullruit?

Barbara Joe: For us, the *cuyakegglit* that have orange tops, they used to have us chew on the membranes of those for a while. And those *ayut* [Labrador tea] on the tundra also, the small tops of Labrador tea that are hard, they used to have us chew on those on the tundra for a while. They say the tops of Labrador tea are also medicine. They also had us drink that kind of tea.

And when we young ones would take baths, they'd have us take baths, they'd tell us to take baths. And after taking baths, they would burn them. They also use them in church today. After drying those pieces of Labrador tea, they would burn them and shake them over our bodies. They say they are *essuircautet* [those used to cleanse and purify]; they are good. They say when we [cleanse] our bodies with *ayut*, we lose our impurities.

Maryann Andrews: Yes, we cleanse ourselves.

Barbara Joe: We cleanse ourselves, we cleanse ourselves.

Maryann Andrews: You know one time we did [a purification ceremony] at Emmonak, when we had some Indians among us, they let us use their kind.

Mary Black: Sage or sweet grass?

Maryann Andrews: I don't know what kind was that, they burned it. When I did that, I really feel it. I did as they told me to.

I could feel it going down and going out down there. I just go like this, to take it away as it went.

After that I feel so light. They're strong those Indians' kind. I feel it. I told Sister Mary, too, Sister Mary here, when we sit down, I ask her, "You feel it too?" She said yaa.

Barbara Joe: It's because they are good. We only feel ones that are good. They say sometimes, some people, I'm not sure how they're living to cause this, they say some people don't feel them. They say only people they consider to be pure, everything [traditional medicine], on a person who is pure, they say only a person who is pure feels that.

Arnaucuaq: Wangkuta cuyakegglit taugaam imkut *orange*-anek kangelget, tamakut qeltaitnek tamuaguraqercetlallruitkut. Imkut cali nunapigmi ayut, ayut kangrit imkut mikcuaraat ayut kangrit tegg'urluteng, tamakunek cali nunapigmi cali tamuagurcetaqluta. Tamakut-llu-gguq yungcautnguut ayut kangrit. Cali yuurqercetaqluta tamakunek.

Cali wangkuta *young*-alriani maqiaqamta, maqivkarluta, ellaita maqisqelluta. Tua-llu maqirraarluta eleggluki. Maa-i *church*-am-ill' aturlarait. Kinercirraarluki tamakut ayut eleggluki qaimtenun evcuarutekluki. Tua-i-gguq essuircautnguut, assilriaruut. Qaivut-gguq wangkutnek ayunek piaqamteki caarrluput kataglaraput.

Tauyaaq: Ii-i, carrirluta.

Arnaucuaq: Carrirluta, wangkutnek carrirluta.

Tauyaaq: *You know one time we did at Imangaq, Indian*-anek ilaluta pillemteni *they let us*, ellaita pimeggnek aturcetellruakut.

Mary: *Sage or sweet grass?*

Tauyaaq: *I don't know what kind was that*, eleggluku. Wiinga tang tua-i pillemni, *I really feel it*. Tua-i alerquallratun, *I did*.

Nallunaunani atrallra kanaggun-llu anluni. *I just go like this*, malikluku ayaucesqelluku.

After that I feel so light. They're kayuq *those Indians' kind. I feel it. I told Sister Mary, too, Sister Mary*-q-llu wani, *when we sit down, I ask her, "You feel it too?" She said yaa.*

Arnaucuaq: Assilriaruata. Assilriit taugaam *feel*-atuaput. Iliini-gguq taugaam yuum iliinek, qaill'-kiq tayima yuungami, *feel*-ayuitait-gguq ilaita yuut. Essuilkemeggnun-gguq taugaam, ca man'a tamarmi, essuilngurmun yugmun, *feel*-atua-gguq tamana essuilnguum yuum.

Maryann Andrews: In the wilderness, even though it's wilderness, there are different kinds of medicine.

Barbara Joe: Yes, the entire surface of the land is medicine.

Wormwood[16]

Paul John: In my village wormwood is called *caiggluk*. And down at Kipnuk, people also call them *caiggluk* since there is someone named Qanganaq [Squirrel] there. And since there's someone named Caiggluk at Eek, [wormwood plants] are called *naunerrluk*. At Quinhagak some also call them *naunerrluk* because there is someone named Caiggluk there. And around here [in the Kuskokwim River area], these are probably called *qanganaruaq* since there is no one named Qanganaq here. I have heard them being called those names. They also call them *naunerrluk*. I have heard them being called *qanganaruaq* here [in Bethel]. In my village they are called *caiggluk*.

They put them in water like they do Labrador tea. Some people like taking a sip of it. The [medicine] works for those who believe that they work. It is helpful. They are good for people who have joint pain, too. Wet them, place them on joints, bind them with bandages, and go to sleep. When my back was hurting, I would do that. I would bind my knee with it and a piece of new fabric and go to sleep. It would be so helpful.

Timothy Myers: I think there are many uses for *caiggluut*. We call them *caiggluut* in our village. When people have sores, too, they dry those [pieces of *caiggluut*] after they pick them, and after drying them, they soften them with a circular motion, and [the *caiggluut*] will become cotton like she mentioned, after drying them. [The *caiggluut*] is then put on the sore on the body. The next day when you check the sore, it draws out the pus along the sore. Then when you put more on, you cover it with that again.

If you check on it the next time, the sores will have already disappeared. In my village since they started taking steam baths, this is what they do. They put [wormwood leaves] in the container of boiling water that is used for pouring onto the rocks [in the steam bath] and let [the leaves] boil in it. Even if [the leaves] are in it, they pour the liquid onto the rocks. It is very good to use, and it seems like [a person] sweats more. These have many uses.

Tauyaaq: Yuilqumi, yuilquungermi camek pitangqertuq iinrumek.

Arnaucuaq: Yaa, yungcautnguuq nunam qainga tamarmi.

Caiggluk wall' qanganaruaq

Kangrilnguq: Nunamteni caiggluugut. Unani-ll' call' Qipnermiut nuniitni Qanganartangqerran call' caiggluuluteng. Iigmi-llu Caigglugtangqerran naunerrluuluteng. Kuinerraami-ll' call' naunerrlugnek ilaita pilarait Caigglugtangqerran tamana. Maani-ll' maa-i qanganaruarullilriit Qanganartailan. Taukuuluki tua niitelaranka. Naunerrlugnek-ll' tua pilaqait. Maani qanganaruanek niitelaranka pivkarluki. Kingunemni taugken tua caigglugnek.

Waten ayucetun mermun piluki. Meq'aquurallrit ilaita assikait. Ukveqestemeggni-ll'-am tua-i nall'arusngaaqluteng. Tua-i ikayuugaqluteng. Usguniqluni-ll' atam assirtut. Mecungluki tua waten, usgunminun ellirraarluki nemerluki qavarluni-llu. Kelugka wii pilallragni pitullruunga waten. Ciisqugka ellivikraarluki nutaranek lumarrayagarmek-ll' nemerluku qavarlua. Ikayuugaqluteng.

Uparquq: Ukut cali caiggluut atullrat amllerrsugnarqut. Wangkuta-ll' caigglugnek pituaput nunamni uani. Cali naurrlugaqata makut kinercirraarluki qang'a-ll' cat naurrluut naumaaqata, naugaqata, avutullratni kinercirraarluki ululriani waten *cotton*-aatun ayuqliqertetuut uum qanrutkellratun kinerciqerraarluki. Tua-i-llu tamakut naurrlugmun elliluku piciatun maavet qaivnun elliluki. Unuaquan paqciiqan imna naurrluk, nucuutelliut naurrluum tamatum pianek. Tua-i-ll' ataam ellivikekuvgu tamakunek patuluku.

Uumikuan paqciiqan imkut naurrluut ak'a tamallrulliut. Tuamta-llu maa-i pilartut nunamni makut maa-i *steam*-atgun maqiyaurcata. Imumun qaltaanun, mer'anun, ekluki qallaucelluki. Tuanlengraata tua ciqiciqiiluni. Assiqapiarluni cali ceq-llu cali tukniriqerrnganani. Caullrat makut amllertut. Waten cali waten nauyullratni neru'urluki, mecuit ig'aqluki cali piyaraugut makut.

These can also be used when they are newly grown by chewing on them and swallowing the saliva.

Marie Myers: In the spring, they are also cooked, even though they withered. Hot water is poured over them, and they are covered like this. Even though they are dead plants, they are useful.

Paul John: I also forgot this a while ago. When my wife and I take a steam bath, she sometimes pours hot water in the basin after putting [*caiggluk*] in it, and she pours the water [with *caiggluk*] on the rocks.

They also swat themselves on their backs or body with them. They have many uses. Since some people believe that they work, it helps them. They use them to swat themselves, *taarriluteng* as they say.

Wassillie B. Evan: Because Ellam Yua created the land, things on the land don't grow by themselves. Only the one who created the world makes them grow. When [plants] die, and when it's spring right after the snow melts, the land appears as though there is nothing on it, the land looks dead, and it seems like they would never grow again. And the trees look like dead, dried wood. God, Ellam Yua created everything living on the land.

And since [the land] is alive, because it was created that way, after there were no longer any berries on it, and even these salmonberries, after they are seen, they fall when they have reached their time to wilt, their limit. When their time comes, other things on the tundra that grow leaves start growing. It is true, Ellam Yua makes everything grow. Because of that, they don't grow by themselves.

Theresa Moses: I never heard of these. We call this part of the low bush cranberries *pellukutait* [their leaves]. These are where they grow. Some people say that they prevent asthma. They say that it helps them with their breathing. When I used to have problems with my breathing, I never took any medications, not knowing about the medicines, only this one.

And when adding it to water that is poured [on rocks] in the steam bath and boiling it and spilling it, [the water] would smell just like Vicks that is around the mouth. It would make the breathing deeper. These are like Vicks.

Luqipataaq: Nalangraata-ll' up'nerkami-ll' ataam pilarait egaluki. Puqlamek kuvluki tua uucetun patuluki. Nalamang'ermeng-ll' atuugut.

Kangrilnguq: Augna avauqalliniaqa. Imutun aiparma-ll' tua-i maqiaqamegnuk iliini ermigcuun uuqnarqellriamek imirraarluku mer'atnek tua-i ciqiaqluku.

Cali-llu taarritekluki patguarcuutekluki tememun tua-i. Atuullrat amllertuq. Ukveqlaamegteki ilait ikayuutekaqluku. Wagg'uq taarriluteng.

Misngalria: Maa-i makut, man'a-wa nunam qainga Ellam Yuanek taqumiimi, cat makut nunam qaingani ellmeggnek nauyuitut. Uum taugaam Ellamek pilillrem nauvkalarai. Tua-i-ll'-am maa-i nalakuneng, up'nerkaqan-llu tua-i, urukan urukarraarqan nuna tua-i tuar cakartaunani, nunam qainga tua tuqumalriatun ayuqluni tuar naunqiggngailnguut. Makut-llu cali napat tuar kinret. Ca tamarmi man' nunam qainga, Agayutem-am, Ellam Yuan taqellrullinia unguvaluku.

Unguviimi-llu, canek atsairuqaarluni-llu tuaten taqellruani, atsairuqaarluni, atsat makut, atsalugpiat-llu maa-i makut tangrruurraarluteng igtelalriit tua-i piunrillerkarteng *limit*-aarteng tekitaqan. Cali-am pinarikuneng, cat allat nunam, nunapiim qaingani naugiluteng *leaves*-aarit naugiarkat. Ilumun tua-i tuaten naugivkarluki Ellam uum Yuan naugivarluki cat. Tamana pitekluku, ellmegnek naunrit'laata.

Ilanaq: Makut-llu niicuitellruanka. Maa-i wangkuta pituaput pellukutait tumagliit. Uitaviit. Ilaita anerniqnaitniluki pilarait. Ikayuuniluki anernemeggnun. Wiinga anernemnek pitullemni iinruyuitellruunga tua cat iinruaqetuutet nalluluki, kiingan man'a.

Cali qasgimun kuvuurutiinun ilakluki qallangutevkarluki kuvelriani tuar ilumun Vicks-aaq man'a-llu Vicks-aacetun qanrem cenii ayuqluni. Man' yuuryiullra-llu call' qamna pivigturiqerrnganani. Makut Vicks-aacetun ayuqluteng.

We did a purification ceremony with smoke[17]

Theresa Moses: Back then they would bring us and give us the tips of the male *ayuq* plants and let us eat them. They took these off and made us eat them. We would chew them ourselves and swallow them. When they ran out of tea, they would drink those [*ayut*] in place of tea. They would also use these for *tarvaq* [cleansing oneself with smoke], and they would say that it took away their sickness, that they are smudging their sickness with it. The tips would be smoking. They would also inhale [the smoke].

And when they put it inside and underneath their garments, they would open this [neck opening of one's garment] and place [the smoking plant] down there [near the garment's hem]. They would stand up and let the smoke go out through [the neck opening]. They would call it *qumigturluni* [putting something inside]. Yes, [the smoke] went inside between one's clothing and one's body, so they had to pull [the neck opening] back and it went out this way.

Paul John: They let their sickness go out.

Theresa Moses: They wanted [their sickness] out and said that this kind [of plant] took out [the sickness]. They burned it. It went out this way, and they smoked their sickness out with this. They called it *qumigturluni*. That is what they did when I watched them.

And once when I watched those who were going seal hunting, I was amazed to see someone light something on fire over there and then go. Then he placed his kayak on the kayak sled, and in a hurry he went over that fire that he had made and left. And then I think I asked my late older sister, "Why is he doing that?" She told me that he was cleansing his kayak before he went seal hunting.

These were medicine from way back then. They were smoke cleansers. They would also put [the smoke] inside their clothes. They let [the smoke] come out that way, and it seems like it was strong.

Paul John: Here is a song of true belief [in using smoke to purify oneself].

May I purify you with smoke, *ayi-yura-nga-yaa-rraa*.

Tarvarlut

Ilanaq: Avani tua-i ak'a ayaulluta-llu makunek maa-i kangritnek angutaita kangritnek nerevkarluta. Makut maa-i aug'arluki nerevkatullruitkut. Wangkutnek tamualuki ig'aqluki. Tua-i caayuutairutaqameng-llu makut maa-i yuurqaqtullrit. Cali tua-i makut maa-i tarvauluteng, ayuurutniaqluteng apqucirrlugteng tarvarniluku. Iquit-wa tua puyiallagalriit. Tua-ll' call' narumaurluki.

Qumigtuquneng-llu call' tamaa-i avukluku una pakigluku kanavet elliluku. Puyur-am waten nangerrluni 'gguun-llu puyur anevkarluku. Tua-i-gguq qumigturluni. Yaa, qamavet iterluni una tua-i pakigluku 'gguun-llu anluni.

Kangrilnguq: Naulluutni anevkarluku.

Ilanaq: Anesqumaluku anetniluku caarrluk, makucimun. Eleggluku. 'Gguun tua-i anluni anutevkarluku tua-i nangtequtni makucimek. Tua-i-gguq qumigturluni. Tuaten cumiksugallemni pitullruut.

Cali qamigaqatalriit cumikeqallemni iillayulqa tua ayagareskili ingna kumarqaarluku, cat tua-i ingkut. Tua-i-llu-q' qamigautegnun qayani qainganun elliqerluku, cukangnaqluni taun' kumartellni qaingirluku ayagarrluni. Tua-llu alqairutka aptellruyugnarqaqa, "Ciin-gguq augna waten pia?" Ava-i-gguq qayani cani qamigaqataami caarrluirai, tarvarai.

Tua makut tuaten avaken ayagluteng makut iinruugut. Tarvauluteng. Qumigturluki tuaten. Tuaggun anevkarluku, tukninganateng cali.

Kangrilnguq: Waniwa ukvekpiarutii yuarun.

Tarvar-nauram-kaan, ayi-yura-nga-yaa-rraa.

May I purify you with smoke, *ayi-yura-nga-yaa-rraa*.

May I brush off defects from your body so you may have renewed strength, *aa-rra-nga yi-yaar, una-rraa*.

Yugi-ama, may I purify your body with the cleansing smoke of Akuluraq [Etolin Strait] down there.

So that you will do well when you go seal hunting, *aa-rra-nga yi-yaarr una-rraa-i*.

May I purify you with smoke, *ayi-yura-nga-yaa-rraa*.

May I brush off defects from your body so you may have renewed strength, *aa-rra-anga yi-yaarr, una-rraa-i*.

Yugi-ama, may I purify your body with the cleansing smoke of Kinguqat [the back side of Nelson Island] up there.

So that you may do well when you gather plants and berries, *aa-rra-anga yi-yaarr, una-rraa-i*.

Labrador tea[18]

Francis Charlie: People always had a supply of Labrador tea. They were their medicine and tea.

Denis Shelden: They also used them to purify with smoke.

Francis Charlie: Today in some churches, they use incense. Our ancestors, too, after lighting Labrador tea, as you know that person there spoke yesterday and said that after leaving our camps for some time, when we would arrive, after [purifying] the inside of the home, we would also [purify] the outside.

A person who cannot see [animals] or cannot catch although he hunts, they have him purify himself with Labrador tea smoke. They have him do that although he wasn't a poor hunter.

Tarvar-nauram-kaan ayi-yura-nga-yaa-rraa.

Elluga-nauram-kaan pikani-rani-aar tuta-ani-aa, aa-rra-nga yi-yaar, una-rraa.

Yugi-ama tarvar-nauram-kaa-rraan Akuluram kat'um tarvartainek-qaa-rraa.

Ellua-rrluten-qaa qamiga-quni-ya-tuta-ani-aa, aa-rra-nga yi-yaarr una-rraa-i.

Tarvar-nauram-kaan, ayi-yura-nga-yaa-rraa.

Elluqa-nauram-kaan pikani-rani-aar-tuta-aniaa, aa-rra-anga yi-yaarr, una-rraa-i.

Yugi-ama tarvar-nauram-ka-rraan, Kinguqat-gguq qamkut tarvartaatnek-qaa-rraa.

Elluarrluten-qaa makiraquni-aar-tuta-ani-aar, aa-rra-anga yi-yaarr, una-rraa-i

Ayuq

Acqaq: Ayuukaraicuitellruut yuut. Iinrukmiluki-llu tua-i yuurqaqluki tuaten.

Kituralria: Tarvarcuutekluki tuaten.

Acqaq: Maa-i *church*-at ilaitni augkunek *incense*-anek pilaqait. Augkut-llu ciuliaput ayunek kumarqaarluki, akwaugaq tua-i tauna qanellrulria unisngarraarluki *camp*-aput tekitaqamta tamakut tua-i qillerrluki enem ilua pirraarluku ataam-llu elatii.

Waten cali yuk-wa tua-i pissungermi tangerciigaculria qang'a-ll' pitesciigaculria tarvarcetaqluku. Tua-i-w' tuaten picuiterrlainanrilengraan.

Alice Rearden: So he purifies himself with smoke?

Francis Charlie: Yes, and he can [purify] himself, too.

Alice Rearden: And his hunting implements?

Francis Charlie: Yes, everything in the home, and he can also purify them, although they're his hunting implements.

Raphael Jimmy: Purifying with Labrador tea smoke starting from long ago, or when I became aware of life, when I observed them, they used it as medicine, and they even used it as tea, as tea. Since my parents constantly ran out of supplies in the fall, my mother would see the amount of tea, of Lipton tea, and if it seemed like they would run out, they would pick these [Labrador tea] and add to it and mix it, and it suddenly increased in amount. They didn't run out. They also used it as medicine. And when I came to observe things, my mother was never without a supply.

Alice Rearden: I also heard that their small tops that are round, that those are also good for those with colds.

Francis Charlie: Yes, they are medicine.

They are what they call the *nasqut* [heads] of the male plants. They refer to those as their *angucaluut* [male plants]. But ones that don't have those [round heads], they call them *arnacaluut* [females].

Raphael Jimmy: Everything, including plants, they knew whether everything on the land, even if it was a plant, was either male or female, anything on the land. They would say that this particular one is a female plant. This is what they say, that their females are softer when they work on them. But the males are hard. They knew what they were. And these grasses have males.

You have to believe in it[19]

Raphael Jimmy: One also does the following. If you eat this, or if you eat the top of [Labrador tea] when you have a sore throat, when you consume it, you have to let it work in your mind.

Cucuaq: Ellminek-qaa tarvarluni?

Acqaq: Ii-i ellminek-llu piyugngaluni.

Cucuaq: Pissurcuutni-llu?

Acqaq: Ii-i, tua-i ca tamalkuan enem iluani, pissurcuutekngermiki-llu piyugngaluni.

Angagaq: Tarvar una avaken ayagluku wiinga-wa tellangellemni tarvar una ayagmek aturaat iinrukluku caayuqluku tuaten yuurqaqluku. Waten-wa angayuqerrluugka uksuarmi nurutelgulaamek, waten aanama tangerrluku un' caayuq, *Lipton tea*-aq imna tangerrluku nurutarkaungataqamek tamakunek makunek, avurluki makut ilaluku *mix*-arluku, amlleriqerrluni-ll' tua-i. Nurucuunatek. Iinruqluku cali tua-i. Murilkua-llu tua-i maaten aanaka piicuunani.

Cucuaq: Augkut-llu niitellruunga kangyagait uivenqeggluteng, tamakut-llu-gguq quselrianun assirtut.

Acqaq: Ii-i iinruugut.

Angucaluit-wa tua-i wagg'u-q' nasqurrit. Tamakut angucaluitnek pilarait. Tuaten taugken piilnguut arnacalunek.

Angagaq: Ca tamarmi naunraungermi piciatun nunam qaingani naunraungraan avani angucaluq arnacaluq-llu nallunritait nunam qaingani piciatun. Una-gguq arnacaluuguq. Waten qanerlartut arnacaluit-gguq atam qetullrulartut waten caliaqaqamegteki. Angucaluit taugken teggluteng. Nalluvkenaki tua-i. Can'get-llu makut angucalurluteng.

Ukvekluku

Angagaq: Waten cali pinaqluni. Uumek wavet nerkuvet waken taumek-llu kangranek igyariqekuvet, atuquvgu umyuarpegun calivkarluku.

Francis Charlie: You have to believe in it.

Raphael Jimmy: You must believe in it. They say it will only work then. But if you just eat it like eating dry fish, it is nothing.

Francis Charlie: It won't work.

Raphael Jimmy: These things on the land, the things that God gave us on the land, we must evidently use them in that way. We must always believe in them, and not think of them as pretend, and we must not think that they have a bad consequence, but a good one.

Alice Rearden: Back when they'd purify with Labrador tea smoke, did they not say anything?

Francis Charlie: They just knew [the reason] in their minds, like this.

Alice Rearden: Someone said, I've heard this before, but don't know who it is, he/she said that some people used to say, "*Ayum ayuarutaa.*"

Francis Charlie: [It's] *ayuurutaa* [taking it away].

Alice Rearden: Taking it away.

Francis Charlie: Yes. They say its smoke was like a shield to animals they hunted, to an animal. Its smoke was like a barrier. If I covered my face, I cannot see this person although he's close. They say that's what they're like.

Alice Rearden: So animals cannot see a person?

Francis Charlie: Yes. When a person purifies himself with Labrador tea smoke, it's like [animals] cannot see him.

Raphael Jimmy: And it also cleanses sickness. . . .

Although one didn't say anything, when I came to observe things, after putting the deceased in a seated position, then when they took [the body] out, then my mother would light a small amount of Labrador tea. Starting from the corner of a home, when they took [the deceased] out of the home,

Acqaq: Ukvekluku.

Angagaq: Ukvekluku. Tuani-gguq taugaam caliciquq. Waten taugaam nernginaarluku neqerrlugtulriatun cauvkenani.

Acqaq: Calivkenani.

Angagaq: Tuaten makut nunam qaingani Agayutem makut cikiutai man'a nunam qainga tuaten atu'urkaulliniavut. Ukvekerrlainarluki, pinguaruvkenaki, assiilngurmek cali iqulirpek'naki assilriamek taugaam.

Cucuaq: Tamaani tarvaraqameng camek qanqayuitellruut?

Acqaq: Tua-i nalluvkenaku taugaam umyuameggnun yaa, waten.

Cucuaq: Augna iliit qanqallruuq, niiteqallrukeka-wa kituuciinaku waten qanlallruniluki ilait, "Ayum ayuarutaa."

Acqaq: Ayuurutaa.

Cucuaq: Ayuurutaa.

Acqaq: Ii-i. Waten-gguq pitarkamun-llu camun ungungssimun cauli capkellriacetun-gguq ayuqluteng pitullruut tuaten puyuan. Tua-i-w' puyua taman' capkellriacetun. Kegginaqa capkumku una tangerciigataqa canimelengraan. Tuaten-gguq ayuqut.

Cucuaq: Ungungssit-qaa yuk tangerciiganaku?

Acqaq: Ii-i. Tua-i-w' tarvaquni tuai tangerciigalngurcetun ayuqluni.

Angagaq: Cali tua-i mat'umun nangtequtmun carriutnguluni. . . .

Qanenrilngermi tua-i maaten murilkua maaten tuqumalria una tuaten aqumlluku waten pirraarluku tua-llu anucatgu aanama-ll' tua-i ayurraat imkut kumarrluki. Waten enem kangiraanek waten ayagluku, anucatgu tua-i man'a ayulluku man'a maliggluku augna anelria kingunrakun.

she would spread Labrador tea smoke in the area, following the one who was going out, following him/her. And she would also put Labrador tea smoke on the small shelves, thinking in her mind, "Sickness, illness just left following that person who is dead." Then when she went outside, she didn't bring it inside but discarded it. That's how my mother used to carry it out.

Francis Charlie: Those I caught in the past used to always do that, like this person said.

Raphael Jimmy: That's how they carried it out. But then, there was no sickness at all after that person did that. And there was no ghostly presence at all. When I observed things, my parents never ever told me a ghost story at all.

Francis Charlie: But when my grandparents would speak, they would say that those who had a bad experience in the place they had come from, those who hadn't had a good experience in their place, it seems they tended to go back to those places. When they do that, they evidently would use Labrador tea as a medicine [to purify the area].

Alice Rearden: When that person started to haunt people?

Francis Charlie: Yes.

Raphael Jimmy: These days, the priests still carry that out. They cannot be without it. They use it to cleanse, and when there are many people at church, they cleanse them and have the smoke take it away.

Francis Charlie: Labrador tea is starting to be used in churches more recently. In the past, they only used incense.

Raphael Jimmy: Incense is *tarvaq* [smoke used for purification]. Although it's incense, since the priest has blessed it, it becomes like *tarvaq*. But today sometimes, when we used to gather at St. Mary's sometimes, you know how they would cleanse us when we would do things, they would cleanse us.

And one has to receive it well. A priest or deacon, if something is wrong with you or if you're sick or if something is wrong with you, if he uses *ayuq* to bless you, then you would accept and receive it as being that. Through the

Qulqitaraat-llu makut ayulluki waten umyuamikun, "Caarrluk nangtequn ava-i augna maliggluku tuqumalria ayagtuq." Tua-llu tellamun anngami, tellamun anngami itrutevkenaku eggluki-llu. Tuaten ayuqluku aanama atutullrua.

Acqaq: Tuaten piuratullruut anguqallrenka-ll' augkut uum qanellratun.

Angagaq: Tuaten atu[tullruat]. Tua-i taugken nangtequkartaunani taum tua-i kingunra caitqapik. Camek-llu alangrurrlugmek-llu caitqapik tang. Murilkua maaten angayuqarrluugka-ll' imkuk alangrumek-llu tang caitqapik qanemcitqallrunrilkiignga caitqap'ik.

Acqaq: Augkut-wa taugaam maurluugka qanerturaqerqamek qantullrulriik kingunerrlugtellriit-gguq taugaam elluarrluteng kingunermeggni pillrunrilnguut tuarpiaq uterrviksulriacetun pitullruit. Tua-i tuaten piaqameng ayunek tua-i iinruaqcarluki pitullrulliniit.

Cucuaq: Tauna alangruungareskan?

Acqaq: Mm-m.

Angagaq: Maa-i cali tua-i tamana agayulirtet aturluku. Piicesciiganateng. Tua-i carriutaqluku mat'umun, yugyagaqata-ll' agayuvigmi tua-i carrirluki iciw' puyuakun ayagcelluku.

Acqaq: Maa-i-w' ukaqvaggun atuurtellria tamana agayuvigni ayuq. Avani *incense*-arrlainarteggun.

Angagaq: *Incense*-aq-wa tua-i tauna tarvaulria. Tua-i *incense*-aungermeng ak'a agayulirtem tua-i asrurtuumiini tua-i uucetun-llu tarvatun ayuqliriluni tua-i. Maa-i taugaam watua iliikun, imumi wangkuta katurtetullemteni imumi *St. Mary's*-aami caaqamta iciw' carritukaitkut, carrirluta tua-i.

Cali tua-i elluarrluku akurturarkauluku. Agayulirtem wall'u-qaa piciatun *deacon*-aam cakuvet nangteqluten qaillun-llu pikuvet ayumek asrurtuqaten tua-i elpet-llu ciuniurluku taunguluku tua-i teguluku. Puyuakun

smoke, your sickness will leave. And you won't recover right away, but you will gradually get better.

Francis Charlie: Only if you believe in it.

Raphael Jimmy: Only if you believe in it.

Francis Charlie: You have to accept it and receive it as being real.

Raphael Jimmy: You receive it.

One who was going down to the ocean for the first time would light a flame[20]

Marie Meade: Miisaq [Frank Andrew] from Kwigillingok said while he was telling stories that they would place some wood down and light them and add some Labrador tea and made it very smoky; then when it was good, he would go through it on his way down with his kayak. His caretakers probably had him do that before he went down to the ocean. Did you also do that in the past?

David Martin: Our elders' customary ways have existed since the time of their ancestors. What you just mentioned, as his ancestors had done, only one who was going down to the ocean for the first time would light a flame, and following the custom of his ancestors, one who was going down to the ocean for the first time would light a fire and like you said, they would light that fire and go to it, and after filling it with Labrador tea, after removing his belt, the one who was going down to the ocean for the first time would place heavy smoke all over his body.

Then after doing that, he would take his kayak and go through the fire and go to the ocean. He would have his kayak go through it, too. They say that if he follows the tradition that was passed on from his ancestors, he will appear bright to the persons of the ocean. Since our elders explained everything we don't know, they said one who was going down to the ocean for the first time is brightening himself so that the persons of the ocean wouldn't be offended by his appearance. You see that our elders explained everything. That one who followed the customary ways of his ancestors.

tamana nangtequten tua-i ayagarkauluni tayima. Egmian-llu wani assiriqertevkenak, pingiinaararluten taugaam assiriinarluten.

Acqaq: Ukvekekuvgu taugaam.

Angagaq: Ukvekekuvgu taugaam.

Acqaq: Ilumun pipiuluku akurturluku.

Angagaq: Akurturluku.

Imarpigmun ayakarraaqatalria kumarciluni kenermek

Arnaq: Miisaq Kuigilngurmiu qanemcillermini pillruuq, cat-gguq elliluki taukut muriit elliluki kumarrluki, ayunek-llu-gguq avuluki, arugpagtelluki; assiriata tumekluki atrarluni, qayani. Imarpik-wa tua-i kanarpailegmi tuaten pivkarluku taukut aulukestain tuaten pillikiit. Tuaten-llu-qaa elpet pilallruuten?

Negaryaq: Ciulialleraput augkut yuut piciryararrarteng tua-i avaken ciuliameggnek ayagluki piciryararraqellrullnikait. Tauna-am tua-i qanemcikellren, ciuliari tamakut tuaten tua-i, imarpigmun taugaam ayakarraaqatalria kumarciluni kenermek, ciuliami augkut atutukiit tua-i tamana nall'arrluku-gguq keneq kumarrluku imarpigmun ayakarraaqatalria, tua-i-ll' ayunek augkunek qanrutkellerpenek keneq tauna kumarqaarluku ullagluku keneq tauna ayunek imirraarluku, naqugutairraarluni iluminun qamavet puyurpagtelluku imarpigmun ayakarraaqatalria.

Tua-i-ll' nutaan tauna tua-i tuatnarraarluni qayani teguluku tauna keneq tumekluku nutaan imarpigmun ayagluni. Qayaminun tamatumun call' tumkevkarluku. Tua-i-gguq tauna-am ciuliaminek tamana piciryaraq atuamiu imarpiim yuinun tanqigcartuq. Nallukek'ngarput ca tamalkuan ciuliamta augkut nalqigcamegteggu, tuavet imarpigmun ayakarraalria tauna tanqigcartuq imarpiim yuinun taukunun tangnerrlukesqevkenani. Tang ca tamalkuan ciuliamta nalqigeskii. Taunaam tua-i ciuliani aturluki piciryararrani tamana atullrullinikii.

Since I never did that, since my elders never did that, I just took my kayak and left. I just left without doing anything. Do you understand?

Those who took plants as gifts[21]

Wassillie B. Evan: Then those two who were going to a dance festival among the *ircenrraat* [other-than-human persons], to that married couple . . . you probably know what *tukut* are; they are the hosts of the place where one stays. They call them by that name when people arrive from different places, *tukut*.

When they arrived there, they say those two people, although they were people, they were white, and their eyes were yellow. They went to those two [hosts]. Then those two told them, "Now, one of you, even if you think those things don't look like anything." . . .

They apparently had sacks made of hide back then, and they placed their things inside them. Then when they danced, since those *ircenrraat* would bring things inside the *qasgiq*, and even though they looked like just grass and other debris, that married couple, the ones with the yellow eyes, those two [hosts] told them, "Take things that the others don't want to take, even though they look like just plants, take some." That's what they told those two [guests].

Then after a while, when they stopped dancing, they did what they were told, the things they picked from down there, and they probably piled them up in the center of the *qasgiq*, as the *ircenrraat* were following the traditions of people, and [the guests] picked whatever they wanted, or I don't think they distribute things similar to the custom of the Eskimos, but picked things themselves, but they picked plants, *pellukutat* [coltsfoot leaves], and they picked other things. They filled it like they had been told to.

When they were done, those two with the light-colored eyes instructed those two who had arrived from outside their area, when they were done, to go toward the area in front of them, and not to go back the way they had come.

Some of the people who were invited had gone back from the place where the *ircenrraat* had their dance festival using their old trail. But they say

Wiinga piyuitellruama ciulianka taukut tuaten piyuitellruata tua-i qayaqa teguluk' ayalalrianga. Qaillukuarpek'nii ayatullruunga tuaten tua-i. Tua-i-qaa taringan?

Carangllugnek tegutellret

Misngalria: Tua-i-ll'-am taukuk ircenrrernun yuraliyak taukugnun-am nulirqellriignun. . . maa-i tukut-wa tua-i nallunrilkeci, tukut; ciunirviit. Tukunek pilarait-am imkut naken tekilluteng, tukut.

Tua-i-am tuavet tekicameng augkuk-gguq yuuk, yuuyaaqeng'ermek qaterrlutek, iikek-llu-gguq cali *yellow*-lutek. Tua-i taukugnun pillinilriik. Tua-i-am taukuk alerqualliniak, "Kitak' ilavtek canek caukenrilkaitnek." . . .

Avani-am kalngangqelalliniut aminek, ca tua-i cat piteng tua-i ekuraqluki. Tua-i-llu-gguq yurallratni tamakut imkut ircenrraat canek taukuk tua-i tuaten pilalliniata canek, itrutaqluteng qasgimun, tua-i-llu-gguq caranglluungalengraata taukuk yuuk nulirqelriik, iilegnek augkuk ecuilnguaraagnek, tua-i-am taukuk alerqualukek, "Ilavtek piyunrilkekainek caranglluungalengraata, avuqitek." Taukuk qanrullukek.

Tua-i nauwa tua-i tayima, tuaten tua-i pisquciatun yuranriata, aruqutetek tuavet unaken qasgimek quyurtelallikait qasgim qukaani ircenrraat maani yuut piciryarait aturluki, canek tua-i piyullermeggnek, wall'u-q' aruqenricugnarqut makut Eskimo-t pilauciitnek taugaam ellmeggnek canek carangllugnek, pellukutanek, canek tua-i avurluteng. Imirluki tuaten pisqutaciatun.

Tua-i-llu qaqiucata imkut, alerqurlukek-am taukuk imkuk ecuilnguaraagnek pilgek [iilgek] avaken tekitellrek, ciunragnegun ayaasqellukek, kingutmun pisqevkenakek.

Tua-i-am ilaita tamakut imkut tumemegteggun kingutmun tuaken ircenrraat yuralriit nuniitnek *invite*-arillret uterrluteng. Taukuk,

those two went beyond that area, going toward the area in front of them, like in the *quliraq*. Yes, like they end the *qulirat* with, "Yes, it is going toward where it's heading, and it is getting straight."

He went that way, and there were a few people who went with him, and by golly, they arrived at their village. But the ones who had returned using the trail that they had used to get there vanished. It was those people.

They say the *ircenrraat* have a doorway up there. A doorway, then they knock, "Come in." [*laughter*] Then the middle doorway, and here is another doorway underneath that goes into the ground. They say those *ircenrraat* have three doorways.

Then when the ones who had left got to their village, they began to have stomach pains. And since they had an *angalkuq* [shaman] among them. . . . The ones who had gone back using their old trail didn't make it home; and they didn't know where they had gone, and didn't arrive. But the ones who had gone in the area beyond them were experiencing bad stomach pains. When they started to have stomach pains, the *angalkut* [shamans] used their powers on them and took out the contents that they ate there among the *ircenrraat*, the different types of plants and debris they had eaten.

Then [the married couple] threw the [skin bags] that they had filled with *pellukutat* [coltsfoot leaves] and plants, and even *qanganaruat* [wormwood, lit., "pretend squirrels"] into the cache and didn't check their contents immediately when they arrived; and then they were there for a while. Then after some time, when her husband went to get something, he saw that skin was stuffed full and was very tight, the one that wasn't too full was stuffed.

Then he opened it, and he saw that those that we used to pick berries with, *pellukutat*, the *pellukutat* that he had placed inside were bearded seal skins. And the *qanganaruat* were squirrel skins. The plants that they had placed inside in the village [of the *ircenrraat*] had transformed into those things.

They had saved those ones who were experiencing stomach pains. Then the ones who had gone [back using their trail] were lost. They arrived the next year.

taukuk-gguq taugken tua-i yaatmun ayaglutek, ciunerteng ciuneqluku, quliracetun. Ii-i, iquklitaqamegteki qalriaciicetun, "Ii-i ciunermikun ayagtuq nakriluni."

Tua-i tuaten ayalliniuq, qavcirrarnek tua-i maligcestai tua-i, aren nunameggnun tekitelliniut. Imkut-gguq-am taugken utertellret tumemegteggun agiirtellerteng pairrluku engelaitut tayim'. Cunaw' taukut tua-i.

Augkut-am imkut amiingqerrnilarait ircenrraat pikani. Amiik, tua-i *knock*, "*Come in*." [*ngel'artut*] Tua-i-llu *middle*-aaq amiik maavet-ll'-am kan'a-wa cali atliq amiik nunam iluanun. Amiingqertut-gguq taukut ircenrraat pingayunek.

Tua-i-llu taukut imkut ayallret engell'icameng ilukaangelliniut, Angalkungqelliniameng-ll'-am imkut. . . . Tua-i yaatmun ayallret kingunitevkenateng natmuruciinateng; tayima tekitevkenateng. Ukut taugaam [ciuneteng] ciuneqluku pillret ilumeggnek aakatanaa. Tua-i-llu tuaten ilumeggnek pingellratni angalkuut tamaa-i tamakut nerellret, ircenrrerni tamaani nerellrit, angalkuita-am qaillun tuunrilluki aqsaqurritnek antaqluki caranglluut ayuqenrilnguut nerellrit.

Tua-i-llu imkuk taukuk imkunek pellukutanek makunek canek tua-i naunraat tuai qanganaruanek tuaten imillrit, ellivigmun tua-i yuvri-rpek'nak' egmian' tekicamek qulvarvigmun milpaulluku tayim'; nauwa tua-i tua-i uitaluteng. Qakuani-gguq maaten uingan cassullermini tangrraa tauna amiq, tenguqlirrluni cagniqapiggluni imna imartussiyaanritler, tenguqlirrluni.

Tua-i-llu-gguq ikirtaa-gguq maaten imkut augkut, iqvarcuutek'lalput, pellukutat, pellukutauluki qemagtellri, naterkaq maklaq. Imkut-llu qanganaruat, qanganarkauluteng tua-i. Tua-i tuani tamakuurrluteng caranglluuluki qemagqaarluki imkut nuniitni pillret.

Tua-llu tua-i tamakut ilukaalriit anirtualuki-am tamakut tua-i pilriit. Tua-llu-gguq imkut yaatmun pillret nauwa tua-i tayima. Tua-llu allamiani nallairtut.

The ones who had arrived there and gone back using their trail, those people would be seen walking around in the sky, and they did things in the sky up there, picking the ice with their ice picks, getting water with their water dippers up in the sky. And when they went fishing, their fish nets would be hanging up in the sky, but those people were paddling up there in the sky. They said those who returned home using the trail they took to get there got stuck up in the sky. I just talked about those three doorways.

They would see those people up in the sky, and they had taken the doorway on top. They didn't arrive when they returned home. But the others [arrived] in their village.

He said the land was thin back then[22]

Cecelia Andrews: To our ancestors, the land was medicine to their bodies. They apparently used it to prevent sickness from afflicting them.

And my grandfather, when he told a story, I was actually small at the time, when I was listening to him; and why was he telling a story about a time when the land was thin? He said the land was thin back then.

He said that he heard about a woman who was gathering greens from the land, and she had a small baby on her back. Then when she crossed through some wetlands, she fell into a wet area.

Theresa Abraham: Poor thing.

Cecelia Andrews: She fell. Gee, she tried to climb out for a long time, trying to cross. As she was trying to step on something solid, the area around her started to rumble, and it was misty. He used to call it *avneruara* [pretend *avneq* ("felt presence of something immaterial")]. It started circling that poor, dear woman, who was in quicksand. They say long ago there used to be a lot of quicksand. Some were very deep. This was before the land fully formed.

Those were evidently around during my grandfather's time. That woman was in quicksand. Poor thing, gee, it so happened that an *avneruaq* was circling her.

Imkut taukut tuavet tekiqaarluteng utertellret, allamiani-gguq yuut
ellakun-gguq tarratnaurtut yuut, caluteng ellami pagaani, tugerluteng,
qaluuriluteng, ellami maani. Kuvyaaqameng-gguq kuvyait agaluteng
amta-llu-gguq ellaggun pagkun anguarluteng. Tua-i-gguq taukut utertellret
ellangqerrutellinii-gguq. Amiignek ava-i pingayunek qanellrulrianga.

Tua-i-llu tamakut tua-i ellakun ellaita tangaagaqluk', amiiget qullirkun
tuaggun pilliniluteng. Ciunitliniluteng-wa tua-i utertellermeggni. Imkut
taugken maavet nunameggnun [ciunilluteng].

Nuna-gguq qecigkitellruuq tamaani

Aluk'aq: Civuliaput imkut, nunaq man'a tememeggnun pilaraat,
yungcautekellrulliniat. Apqucimun tua-i agturyailkutekluku.

Taun-llu ap'aka call' qanemcillrani miktellruyaaqua, niicugniurallemni;
ciin-llu nuna qecigkitellranek qanemcia? Nuna-gguq qecigkitellruuq
tamaani.

Augna-gguq arnaq niiskii makiraluni tunumiayagarluni piipimineng. Tava-
llu-gguq caarrlugkun qeryaryaaqellermini iggluni merrlugmun.

Paniliar: Nakleng.

Aluk'aq: Iggluni. Ala-i mayungnaqumalun' qerangnaqluni.
Tut'engnaqnginanrani-gguq man'a tem'irtuq avatiini, minegluni-gguq.
Avneruaneng pitullrukii. Uyivaa, uyivaangaa-gguq tauna arnaurluq,
aangiitellria. Aangaat-gguq amllepiatullruut akaurtuq. Et'upiarluteng ilait.
Nuna nauluaqerpailgan.

Ap'ama tamakut nalliini uitallrullinilriit. Aangiilluni taun' arnaq.
Nakleng-gguq tang, ala, cunaw' avneruam uyivaarluk'.

Ruth Jimmie: What is an *avneruaq*?

Cecelia Andrews: I'm not sure what those are. They say it has mist, it has some mist.

Theresa Abraham: How scary.

Cecelia Andrews: They used to call them *avneruat*. They are some kind of apparations that make noise.

Ruth Jimmie: So did that one survive?

Cecelia Andrews: I'm not sure what happened to her, she probably did survive.

Ruth Jimmie: Probably because she survived, she had a story to tell.

Angalgaq: Cauga avneruaq?

Aluk'aq: Caugaqat tamakut. Minegluni-gguq tua-i minengqerrluni.

Paniliar: Ala-i.

Aluk'aq: Avneruaneng acitulqait. Em'emiyagtellriit cat alangrut.

Angalgaq: Tua-llu-q' anagtuq tauna?

Aluk'aq: Qaillun-wa tua-i tayim' taun' pia, anallrullilria.

Angalgaq: Tua-i-w' anallrulliami qanemcikangqellinilria.

UUGNARAAT NEQAUTAIT

MOUSE FOODS

Grass seeds and tubers of tall cottongrass[23]

Pauline Jimmie: And [we gathered] *utngungssaraat* [grass seeds], but *anlleret* [the tubers of tall cottongrass] evidently rot. But these *utngungssaraat* don't rot and also these *negaasget* [silverweed tubers] [don't rot]. They are very delicious in broth soup and taste good in seal meat soup.

Martina Wasili: They also stored *anlleret* by digging the ground down to the permafrost.

Pauline Jimmie: And *anlleret* are delicious in *akutaq*. They always tried to make *akutaq* in the past. And with newly-frozen ice at freeze-up, after we picked berries in the fall, right after freeze-up, she would crush the ice, after she melted those *kaugat* [pieces of lard]. What do they call them in English? Tallow. She added that, and she used ice chips as [a substitute for] snow and made *akutaq*. But they weren't bad-tasting; they tasted good. And they also added caribou back fat and made *akutaq* with seal oil in it.

Some time ago at my uncle's home, I ate that kind of *akutaq* with caribou back fat mixed with seal oil. I ate for a long time.

Gathering mouse food[24]

Bob Aloysius: And in the fall, when it began to get cold like this, they would bring us to islands and tell us to run around along the shore. While running along the shore sometimes, our leg would suddenly sink in. They'd suddenly get happy and quickly come to us, and they'd pry open the ground, and there would be many things about this long, and there were even *qet'get* [root nodules of horsetails] with them. Those things that the poor voles worked so hard to gather, we'd steal them. [*laughter*]

They'd tell us to replace it with something small, and even a small piece of dry fish, showing gratitude for voles gathering those. And some [vole caches] had grass in them. You know what *qet'get* are, right?

Alice Rearden: Those [that look] like blackberries.

Bob Aloysius: They are small black things. Yes, it's like they have milk inside them, mm-m.

Utngungssaraat anlleret-llu

Kangrilnguq: Utngungssarnek-llu call' makunek cali, anlleret taugaam makut arutulliniut. Utngungssaraat taugken makut aruyuunateng negaasget-llu. Yuurqaarnun askacagaraqluteng taqukanun-llu keniranun assiraqluteng.

Cuyanguyak: Augkut-llu anllerucitullruut nuna elakarluku kumlii tekilluku.

Kangrilnguq: Akutanun-llu aspiat anlleret akutaqaiceteng. Akutengnaqu'urlulallrulriit-wa. Cikuqamek-llu cikuqerqan aanaka im' uksuarmi waten iqvaqurraarluta waten cikuqaan cikut ciamlluki, egcetaarraarluni augkunek imkunek kauganek. Kass'atun canek pilartatki? Tal'u [*tallow*]. Yaa, tamakunek avuluku, cikunek qanikcirluku akutaqluni. Cayuunateng taugken; assiraqluteng. Tuntut-llu tunuitnek cali avuluki call' akutaqluteng uqumek.

Qangvaq im' angalleramni nerellruunga tamakucimek tunulegmek uqumek avuluku. Nerumalrianga.

Pakissaagyaraq

Sliksuuyar: Uksuarmi-llu waten nenglliryaurtellrani qikertanun ayautelallruitkut cenami tamaani aqvaquasqelluta. Ilaitni aqvaquanginanemteni murullagnaurtukut. Quyallagnaurtut, taigarrluteng pakignaurait ik'iki tamakut waten taktaurluteng, augkunek-llu qetegnek avuluteng. Avelngaat cakviurutkeurlulallrit wangkuta tua-i teglegaqluki. [*ngel'artut*]

Qanrutnauraitkut carraungraan neqerrluggarmek-llu wall'u-q' cimiqaasqelluku, quyavikluki avelngat tamakunek quyurcillratnek. Canegnek-llu ilait avungqelallruut. Qet'get nallunritaten?

Cucuaq: Augkut tan'gerpagtun [ayuqellriit].

Sliksuuyar: Tunguyaarluteng. Ii-i, iluit-llu tuar mulukuungqellriit, mm-m.

They used to have us gather those during the fall before freeze-up, telling us to pick those because they never spoil.

Red and black berries never spoil; you keep them all winter. And then they're really good. You have red, black, *elagaq* [alpine sweet vetch, known as "Eskimo potatoes"] and *qetek* [horsetail root nodule] *akutaq*, mixed, man, man! No sugar, no shortening and sugar. Whitefish, pike, a little bit shortening or sheefish oil to make it fluffy, and you blend them in.

Nick Andrew: He's going to make us hungry.

Bob Aloysius: The things that you were curious about, the subsistence way of life, the entire land can provide every kind of food to us. That's why our ancestors were always eager to gather things from the land, from in the land, and the water. And [they'd subsist] every day.

Gathering mouse food in spring[25]

David Martin: Also during that time, while I was at spring camp while I was a boy, my late mother, after putting on my skin boots in the morning, when she finished, she handed me my small bow and would tell me, "Go out to the areas where the snow has melted again." We were fortunate that there were many voles during that time, "When you see voles, shoot arrows at them, and when you kill them, don't discard them, but bring them to me right away." That's what she would instruct me to do.

One day, just as I had been told to, I went to an area where the snow had melted from the ground, when the snow began to melt. I got to the end, and they appeared like a group of ptarmigan pellets that were partially melted, and there were many of them before me. I took one and broke it in half. It was an *anllereq* [tall cottongrass tuber]. It was mouse food that they had gathered in a cache that they had taken out, and there were many of them along the edge of a place where the snow had melted from the ground.

I took some of them and since our home was nearby, I went to my mother and showed her. I told her, "I found these back there, and I thought that they were ptarmigan pellets since there were so many of them, and I checked one and saw that it was a tall cottongrass tuber." [She said,] "Gee,

Tamakunek tua-i uksuarmi cikuvailgan cumigtevkalallruitkut
iqvaasqelluta tamakunek *because they never spoil.*

Red and black berries never spoil; you keep 'em all winter. Tua-i, and they're really
[good]. You have red, black, elagaq and qetek akutaq, mixed, man, man! No
sugar, no shortening and sugar. Whitefish, pike, little bit shortening or sheefish oil
to make it fluffy, and you blend them in.

Apirtaq: Kaingevkaqataraakut.

Sliksuuyar: Tamakut tua-i paqnayulci nutaan, nerangnaqsaraq,
nunam tamarmi cikiryugngaakut neqkanek. Taumek augkut ciuliaput
cumigterrlainalriit nunam qainganek, nunam iluanek, mermek.
Unuaquaqan-llu tua-i.

Pakissaagyaraq up'nerkami

Negaryaq: Tamaani-am cali tua-i yaani up'nerkiyarani pillemni
tan'gaurlullraulua, aanallma aug'um unuakumi piluggaarlua taqngamia
urluvcuaqa tailluku pilaraanga, "Kitak' qagaani urunquni atam."
Uugarnaat anirta tuani amllellruut, "Ugnarnek tangrraqavet pitgarraarluk'
tuqutaqavki pegtevkenaki taigartaqluki piniaten." Tuaten alerquraqlua.

Pivakarlua-am tua-i tuaten pisqumalqa taman' aturluku, urunquq una
ullagluku urungellrani qanikcaq. Iqua una tekicarturqa, wangni ukut
aqesgirugaat anait ukut urugaarluki, tua-i amllerrluteng wani. Iliitnek
teguqerluku kep'artaqa. Anllerullinilria tauna. Cunaw' waniw' uugnaraat
antait taukut, amllekacaarluteng urunqim iquani.

Ilaitnek-am teguqaullua yaa-i nev'ut canimelan ullagluku aanallemnun
taukunun maniaqa. "Ukunek tang nataqutellrianga piani aqesgit
anaqsuksaaqellemnek amllessiyaagata, maaten tang iliit piaqa
anllerullinilira." "Aren patagmek aqvanaupuk." Ca tauna ciuneqa

let's go get them right away." The one whom I went to quickly got ready, "Let's go get them right away." We went to get those. She filled that large woven grass basket and would say, "Oh my, you found a lot of food to eat."

That's what happened to me back when we were going through a shortage of food, I found those in an area along the edge of where the snow had melted. It was mouse food that they had gathered in a cache. As we are living, when we have lost our sense of humanity, the unseen entity causes us to go through hardship in order for us to come to our senses.

Mice are like people[26]

Theresa Abraham: Mouse food. Yes, those unappetizing ones are among them.

I'm not sure what they call them, but we don't gather those since they don't taste good.

We pick *negaasget* [silverweed tubers] and *utngungssaraat* [grass seeds] down on the marshland. And we leave those that don't taste good alone. . . .

Cecelia Andrews: And when we are sick, we could probably even have the juice of that *neqnialquq* [bad tasting thing], filling a spoonful and eating them. Gee, since they're probably medicine, I was thinking that. Mice won't mistake things [for food] and gather them. I think they were showing us to use those for something.

Ruth Jimmie: They wanted us to eat them along with other things. [*laughter*]

John Andrew: Back when I didn't know things, after looking at [mouse caches] that were filled with all sorts of things, I would sometimes cover them again. I would think, "Those who are lazy probably gather a mixture of things." They gather what they could eat.

But some of them had very clean [caches] that were one type [of food] that they gathered.

upngartuq, "Patagmek aqvanaupuk." Tua-i-am aqvaluki taukut. Issrallugpayagaq augna imiqii aren qanraqluni, "Aling neqkarugarnek ukunek nataqutliniuten."

Tua-i tuani-am tuatnallruunga wiinga nurusngallemta nalliini, taukunek nalaqullua, urunqiim iquani. Cunaw' uugnaraat antait taukut. Ilumun tangvaumanrilnguum ellangcatuyaaqelliniakut tua-i pivakarluta, ellairutsiyaagqamta.

Uugnaraat yugcetun ayuqut

Paniliar: Uugnaraat neqait. Ii-i, avungqetuut tamakunek neqnialngurnek.

Canek-wa pilartatki taugaam neqnialata tamakut piyuitaput.

Augkut negaasget utngungssaraat avurluki tua-i pilaraput unani marami. Tamakut-llu neqnialnguut uitalluki.

Aluk'aq: Nangteqkumta-ll', naliak, mecuaneng taum neqnialquum, *spoon*-acuaraat-llu imirluki neryuumaukut. Ala-i *medicine*-aungata-wa umyuarteqngartua. Uugnaraat alarqelluteng caneng ang'arqeng[aitut]. Maniitetuyaaqellrungataatkut cakesqelluki tamakut.

Angalgaq: Avukeqtaaruk neresqelluki. [*ngelaq'ertuq*]

Alegyuk: Wiinga nallumallemni tamaani piciatun imalget tangrraarluki ilait ataam patutelallruanka. Umyuarteqaqlua, "Qessanquulriit-wa avukuulluki pilallilriit." Neryugngakmeggnek tua-i quyurtelalriit.

Ilait taugaam carrinqepiarnek ataucinek kiingan avuraqluteng.

Theresa Abraham: Yes.

Ruth Jimmie: These would be like the mouse version of *tegaq* [male ringed seal in rut, smelly and inedible]. It's not tasty.

Cecelia Andrews: I learned those, too. They said *utngungssat* were ringed seals. And they said these, those *marallat* [silverweed tubers], the long, small potatoes, they'd call them *makliit* [adult bearded seals]. And they said the ones that were the most unappetizing, they called them *tegat* [male ringed seals in rut].

That's what they used to call those, that mice see them as those.

Ruth Jimmie: So that's how it is.

Cecelia Andrews: They also say they are *ircenrraat* [other-than-human persons]. They say they are a certain way.

Ruth Jimmie: I'm grateful to learn that.

Cecelia Andrews: Yes, I used to reveal the things that I know.

Theresa Abraham: Mice are evidently good. . . .

I don't know about that either, but I was just thinking that like us people, we don't mix our foods. When this type of food is different, we put them in separate containers. Since they know about them, they probably divide those foods like that just as we do.

Ruth Jimmie: Indeed. When they tell their children, "Go and get *utngungssaraat*." They'd know what they are. [*laughter*]

Theresa Abraham: Since they evidently see them [like that], they see these [mouse foods] as something, they see these as something.

Ruth Jimmie: Yes, how nice.

Cecelia Andrews: And I used to hear that when those mice prepared food,

Paniliar: Ii-i.

Angalgaq: Makut-gguq im'ucetun ayuqeciqut uugnaraat *version*-aarit, tegaq. Neqnianani.

Aluk'aq: Wii-ll' elitelqanka tamakut. Utngungssat nayiuniluki makut-llu-gguq imkut augkut marallat takelriit *potato*-yagaat maklagneng pinaurait. Tamakut-llu-gguq neqnialqut teganeng.

Taukut aptulqait tavaten, apertull', ellaita-gguq tua-i taukut uugnaraat tamakuuluki pituit.

Angalgaq: Cunawa.

Aluk'aq: Ircenrrautuniluki cali. Caneng qaill' ayuqetuniluki.

Angalgaq: Quyan' nallunriqerrluku.

Aluk'aq: Ii-i, nallunritellrenka wii aperqutulqanka.

Paniliar: Uugnaraat assilliniut. . . .

Wiinga-llu uunguciicaaqaqa taugaam umyuartequa ava-i wangkucicetun yugcetun neqput maa-i avukulluki piyuilkevut. Ukut all'augaqata allakaita caquluki. Tuaten ellaita nallunrilamegteki neqautek'larngatait wangkucicetun.

Angalgaq: Ilumun. Ellimerikuneng irniameggnek, "Utngungssaarnek aqvatqaa." Nalluvkenaki. [*ngel'artut*]

Paniliar: Ava-i tangtulliniata, cauluki ukut tangerrluki, ukut cauluki tangerrluki.

Angalgaq: Yaa, assirpaa.

Aluk'aq: Tuamtellu niitetullrulua tamakut-gguq uugnaraat uptaqameng

close to winter, a lazy [mouse], they said they would tell them to gather those. They said when they were lazy, after running after them, they'd do something to their children. They would tell them [to gather], and when they wouldn't listen, they would also bite them.

Ruth Jimmie: They proabably also kill them.

Cecelia Andrews: Yes. They said they would gather around him and really scold that one, telling him not to be lazy.

Ruth Jimmie: Like people.

neqkaneng uksurniararaqan qessamkilria ellimerrinaurtut-gguq
ang'arcesqelluki tamakunun. Qessaaqata-gguq malirqerraarluki canaurait
tamakut irniateng. Ellimerrluki niicunritaqata-ll' caqluki cali.

Angalgaq: Tuqutelaryugnarqait cali.

Aluk'aq: Ii-i, Imkurluk' quyurulluku-gguq nunurnauraat cakneq,
qessasqevkenaku.

Angalgaq: Yugcetun.

IQVARYARAQ
BERRY PICKING

> Woman and child pick berries on the tundra near Chefornak, early 1960s.
UNIVERSITY OF DELAWARE, UNIVERSITY MUSEUMS, MABEL AND HARLEY
MCKEAGUE ALASKAN INUIT COLLECTION

Berry picking[27]

Theresa Abraham: And again, one time when I happened to take notice of my surroundings, [my mother] was twining some wheat grass. We had started to go and pick crowberries when she asked me to come with her, and she'd go along the mouth of Cevv'arneq [Urrsukvaaq] River near the old graves and pick crowberries.

Since there was nowhere to store them, and since they had no freezers in the past to store them, since these crowberries tend to turn a light color when they are kept on the ground. Inside a *naparcilluk* [twined grass basket], starting along the bottom, she placed sour dock inside there to line it, and she filled it with those blackberries. I think they also do that with salmonberries.

As she continued on up from the bottom, lining the twined *naparcilluk* with sour dock, she would periodically fill it [with berries]. And when it was filled, she would also secure the top so that blackberries wouldn't come out, and she'd sink it down in the water along the edge of a lake, and it was kept underwater.

One time I told my mother, "The berries we picked." Although I didn't pick berries too well, I was concerned that the berries we picked would spoil. She said that they wouldn't spoil, that I would see them in their original state and not scattered.

I was curious when she was just about to pull them out of the water. And when it was time, probably when she felt it was the right time, during the fall she pulled it out of the water, just when the lake was about to freeze, but probably before the ice became thick. I looked at them out of curiosity, and I saw that the contents were good. They weren't light in color at all, and they weren't scattered everywhere.

These days, if we run out of space in our freezers, we could place those berries inside those [baskets underwater], by following what we observed in the past. That's it.

Iqvaryaraq

Paniliar: Tuamtellu cali, cumikarrlua-am, qayikvayagnek tamaa-i cali tupigluni. Tamaa-i tan'gerpagtelangarrlunuk tua-i cali unayaqaqanga, Cevv'arnerem-llu painganun qungullret augkut nuniitnun tan'gerpagcuraqluni.

Tua-i tayima natmun pivigkaitelaameng, *freezer*-aanek-llu makunek pivigkaitelaameng, qat'rituata-ll' makut tan'gerpiit uitauraraqameng nunami qat'rilartut tan'gerpiit. Tuaten tua-i tauna naparcilluk, quagcinek makunek kanaken ayagluku quagcit imumek elliurturciqai, tua-i-ll' tamaaken tan'gerpagnek imirturluku. Naunrat-llu pilarngat'lallruit cali.

Mayullra maligqurluku quagcinek *line*-irturluku tauna canek tupigaq naparcilluk imirturciqaa. Tua-i imangekan painga cali maqsugnairqaarluku, nanvam uuggun ceniinun kic'elluku, tua-i kisngaluni tayima.

Caqerluku piaqa aanaka, "Iqvapuk taukut." Wiinga elluarrlua iqvaqsaicaaqelrianga, iqvapuk taukut ikiurrnayukluki. Qang'a-gguq ikiurrngaitut, tangerciqanka-gguq ayuquciicetun peksagtevkenaki-llu.

Paqnayuglua-am tua-i nugteqatallrani. Tua-i-llu pinarian tayima cuqni pillian uksuaran nuggluku, tauna cikuqatarartelluku mamturivailgan taugaam pillikii. Maaten tua-i paqnakaqa imai tua-i assirluteng. Qat'riyugnaunateng, peksagcugnaunateng-llu.

Maa-i wangkuta tayima kumlivigmi-llu cipcikumta camek, tamakunek naunrarrlugnek ekviksugngayaaqaput tayima taman' aturluku tangvalalput. Tua-i.

Ircenrraat are indeed very real[28]

Alice Mark: I thought these *ircenrraat* [other-than-human persons] were indeed real. In the past, I'd heard of but had not seen berries that had been picked by *ircenrraat*. They say they pick berries that are very close together. Since I saw it, I finally believed in it. And we were given the following instruction, "Don't take too many, but take only one."

We saw the berries picked by *ircenrraat* one of the many times that we went berry picking. Sammy quickly came over to us, and he said that he saw berries that were very close together over there. Then they started to say, "Maybe these are berries picked by *ircenrraat*." Then we all went over to look at them. We really lacked berries to pick at the time, we really lacked berries to pick at the time, and the berries were scarce.

Then I recalled that instruction to take just one [berry]. When we went to the [berries] Sammy had found, they were gathered and they appeared like someone had put them there, berries that were gathered close together.

I think they were as much as the contents of a cup. They looked like a person had just put them there. I finally believed when I saw them.

Since I heard about it, I told the others with me, after first taking one myself and putting it inside my bucket, "Now, each of you take just one." They each took just one. [*laughter*]

Then [someone] took all of them. [*laughter*] He took all the ones that were gathered together. Then after there had been hardly any berries, there were many berries. And our [buckets] were filled right away. We got lots of berries. I hear that [*ircenrraat*] give [people] berries. They say we should take just one.

We picked berries and [the one who took all the berries] was just walking around up there. When he came, he said that his bucket didn't get filled. [*laughs*]

[He] took all the [berries] that were gathered together. We laughed for a long time. His bucket was empty.

I was thinking, "That [teaching] is evidently true." They say if we come

Ircenrraat piciuqapigtut

Inaqaq: Ircenraat makut ilumun piciuyukluki umyuarteqellruunga. Niitelartua, ircenrraat iqvaitnek tangeqsaitua. Iqvatuut-gguq quyungqapianek. Wiinga tua-i tangellruamku nutaan ukvellruunga. Waten-llu qanrucimalalput, "Amllerissiyaagpek'naci-at ataucirrarmek teguskici."

Ircenrraat iqvaitnek tangellruukut atsiyarpakarluta. Sammy taigartuq, yaani-gguq atsanek quyungqapianek tangertuq. Tua-i-ll' qanngartut, "Ukut ircenrraam iqvaqlii." Malikluta-ll' paqluki. Atsaillitapiarluta, atsaillitapiarluta avacaaraqapiggluta.

Tua-i-ll' tamana wii umyuaqerrluku ataucimek tegutarkaq. Maaten tua-i ullagavut Sammy-m nataqutai quyungqalriit tuartang yuum elliqallri quyungqapiat atsat.

Saskam imaqellii-llu. Yuum tuar elliqallri. Nutaan ukvellruunga tuan' tangrramni.

Tua-i-ll' niitelaama ilanka ukut pianka, wii ciumek pirraarlua ataucimek tegullua qaltaq imiqerluku, "Kitek elpeci-llu ataucitaarluci teguci." Ataucitaarluteng tegulluteng. [*ngel'artut*]

Tua-i-ll' ukut [kia] pingraata tamalkuita teguluki. [*ngel'artut*] Quyungqalriit tamak'acagaita teguluki. Tua-i-ll' maaten atsaitellrurraartelluk' atsarugaat. Egmianun-llu muiriluta. Atsalipiarluta. Cikituniluk' niitelartua. Ataucirrarmek-gguq teguskilta.

Tua-i iqvarluta, [kina]-wa tarrartellria pavani. Taingami qanertuq elliin-gguq tua-i qaltaa imangenrilnguq. [*ngel'artuq*]

Tamalkuita tegullrui quyungqalriit. Ngelaumalriakut. Qaltaa imaunani.

Umyuarteqlua tua-i, "Ilumuulliniuq tamana." Atsanek tekiskumta

upon some berries gathered in one spot, we should all take just one. When I recalled that, I had the people there [each take one].

He came upon some cranes[29]

Katie Jenkins: [Pike] couldn't enter. He left, he left and after a while, there were many cranes. He came upon some cranes. The cranes were jumping up and down, they were extremely happy.

"Oh, how there are no eyes for them to have."

"Oh, how we don't have eyes."

Then one of them took a crowberry. He put them on, but gee, what the crane saw was very dark. Then [he said], "Gee, my eyesight is worse than before."

Then he went along and saw some cranberries. Once again, he put some cranberries in his eyes. When he looked around, what he saw was very red! [*laughter*]

"These are too bad; they apparently aren't good to use since they are red."

He once again replaced his eyes with some blueberries. After putting on those eyes, he looked around, and gee, what he saw was very good. [*laughter*]

Those two became his eyes.

Marie Alexie: That's why their eyes look like blueberries.

Marie Meade: Yes. That one you heard was Aatacuar's story?

Katie Jenkins: One I heard, one I heard.

quyungqalrianek ataucitaagluta tegucesqelluta. Umyuaqngamku
pivkallruanka augkut.

Qucillgarnun tekilluni

Uruvak: Iterciiganani. Tua-i ayaglun', ayaglun'-am piqertuq qucillgarugaat
imkut. Qucillgarnun tekilluni. Qucillgaat-gguq qeckartaalriit,
arenqianateng angnirluteng.

"Iikaipagtat."

"Iingipagceta."

Tua-i-ll' iliita tan'gerpagmek tegullun'. Acaaqak', aren-gguq tangellra
tungurpak qucillgaam. Tua-i-ll', "Aren tangerciigaliirtevsiartua."

Tua-i-ll' ayagluni piuq-gguq kavirlit. Kavirlignek-am call' iililliniluni.
Tangertuq-gguq tangellra kavirpak! [*ngelaq'ertuq*]

"Ukuk wan' arenqiatuk; atuyunaitelliniuk kavircelamek."

Tua-i-am iilinqigtelliniluni call' suragnek. Iilirraarluni kiartuq, aren
nutaan-gguq tangellra assiqapiggluni. [*ngel'artut*]

Iiksagullukek tua-i.

Akalleq: Taumek iingit surarngatellriit.

Arnaq: Ii-i. Tua-i taum Aatacuaraam qanemcia tauna niitellren?

Uruvak: Niitelqa tua-i, niitelqa.

NAUCETAAT
PLANTS

Mercuqelugaat (sea lovage) growing along the coast near Quinhagak, June 16, 2018. JACQUELINE CLEVELAND

Note: In the Yup'ik language, nouns ending in "q" are singular, nouns ending in "t" are plural, and dual nouns end in "k."

In this list, places where particular plant names are used are indicated as follows: (Y) for Yukon; (K) for Kuskokwim; (NS) for Norton Sound; (C) for Canineq (lower Kuskokwim coastal area); (HBC) for Hooper Bay/Chevak; (NI) for Nelson Island; (Akula) for the Akulmiut area west of Bethel, and (N) for Nunivak.

NAUTULIT NERTUKNGAIT
EDIBLE PLANTS

....

allngiguaq/allngiguat	marsh marigold/s (lit., "pretend *allngik* (boot sole patch)"); also *allmaguat* (HBC), *irunguat* (Y) (lit., "pretend *irut* [legs]"), *uivlut* (N) (from *uive-*, "to circle," because of their round shape)
angukaq/angukat	wild rhubarb/s (from *angun*, "male"); also *nakaaret, arnaurluut* (lit., "poor dear *arnat* [women]")
cetuguaq/cetuguat	fiddlehead fern/s, wood fern/s (lit., "pretend *cetuk* [fingernail]"); also *cetugpaguat* (lit., "pretend long fingernails") and *ceturqaaraat* (from *cetur-*, "stretching one's legs)
ciruneruaq/ciruneruat	kidney lichen/s (lit., "pretend antlers")
ciutnguaq/ciutnguat	saxifrage (lit., "pretend ears")
elagaq/elagat (K)	alpine sweet vetch, "Eskimo potatoes" (lit., "things dug from underground"); also *qerqat* (Y)
elquat epuit	rockweed (lit., "stem for *elquat* [herring eggs]); also *qelquat*
iitaq/iitaat	tall cottongrass plant/s, especially the edible lower stems
it'garalek/it'garalget	beach green/s (lit., "ones having little feet"); also *itegarat* (from *itegaq*, another word for foot), *qelquayak* (from *qelquaq*,

	"rockweed, kelp"), and *tukulleggaq* (from *tukullek*, the Cup'ig (Nunivak) word for foot)
kapuukaq/kapuukaraat	buttercup/s (from *kapur-*, "to poke or stab"); also *qaqacuqunat* (Akula) (possibly from *qaqaq*, "red-throated loon")
kun'aq/kun'at	edible root/s of spreading wood fern
mecuqelugaq/mecuqelugaat	sea lovage; also *mecurtulit, mecuqelluqaat,* and *mecuggluggaq* (all deriving from *mecuq*, "liquid part of something, sap, or juice"), and *mercurtulgaat* or *mercurtuliaraat* (from *mer-*, "to drink")
*mecurtuliaq/*mecurtuliaraat	saxifrage; also *mecugtuliaq* (from *mecuq*, "liquid part of something, sap, or juice")
muugarliar/muugarliarneret	root/s of *nuyaruat* (water weeds, lit., "pretend hairs")
nasqupaguaq/nasqupaguat	nutty saw-wort (lit., "pretend *nasquq* [head]")
neqnirliar/neqnirliaraat	oysterleaf/s, beach bluebell/s (lit., "best tasting things")
palurutaq/palurutat	edible mushroom/s (possibly from *palurte-*, turned over or belly-down)
panayulit neqait	bluebells (lit., "bumblebee food"); also *kulukuunaruat* (lit., "pretend bells")
pingayunelgen/pingayunelget	marsh fivefinger plant/s (from *pingayunlegen*, "eight"); also *mecungyuilnguq* (from *mecuq*, "liquid part of something, sap, or juice")
qatlinaq/qatlinat	stinging nettle/s (lit., "those that sting"); also *qacelpiit* (lit., "those that really sting"), *qacellinat* (lit., "those that sting"), *mingqutnguayaalget* (lit., "ones with small *mingqutet* [needles]")
qecigpak/qecigpiit (NI)	unidentified aquatic plant growing in saturated ground along mountains, floating in water (lit., "ones with big *qecik* [skin]"); also *qeciguut* (lit., "big skinned things"), *qecigtut* (lit., "ones with thick skin")
quagciq/quagcit	sour dock; also *aatunat, quunarlit* (from *quunarqe-*, "to taste sour")
quunarliaraq/quunarliaraat (NI)	mountain sorrel (lit., "small *quunarlit* [sour dock]")

taqukanguaq/taqukanguat	reindeer lichen/s (lit., "pretend *taqukaq* [brown bear or seal]"); also *tuntut neqait* (lit., "reindeer food"), *ungagat* (lit., "pretend *ungiit* [whiskers]"), *qilungayagaat* (from *qilu*, "intestines, entrails")
tarnaq/tarnat	cow parsnip/s (*tarnaq* also translates as soul or spirit)
tayarum iluraa/tayarut ilurait	mare's tail plant/s, *Hippuris vulgaris* (lit., "mare's tail's cousin")
tayaruq/tayarut	mare's tail plant/s, *Hippuris tetraphylla*; also ?*tayarulungniit*
tuutaruaq/tuutaruat	rose hip/s (lit., "imitation labret")
ulevleruyak/ulevleruyiit	woolly lousewort
ulqiq/ulqit	tuberous spring beauty, Eskimo potato/es; also *atkallaat* (NI)
uruq/urut	moss/es, Sphagnum moss/es
urutvaguaq/urutvaguat	variety of moss, possibly juniper polytrichum moss (lit., "big pretend *urut* [mosses]")

NAUCETAAT IINRUKTUKNGAIT
MEDICINAL PLANTS

agyam anaa/agyat anait	puffball fungus (lit., "star feces")
anuqetuliar/anuqetuliaraat	yarrow (from *anuqa*, "wind"); also *anguqetussngit* (K)
arakaq/arakat	birch-bracket fungus or "punk" (lit., "ones that will be turned into *araq* [ash]"); also *kumakaq* (from *kuma-*, "to be lit or burning")
araq	wood ash; also *qamlleq* (from *qame-*, "to die down [of fire]")
atsaruaq/atsaruat	wild chamomile, pineapple weed (lit., "pretend *atsat* [berries]"); also *itemkeciyaat* (from *itemkar-*, "to kick lightly"), *kitengkaciyaaret* (from *kitengpag-*, "to kick hard"), *itegmigcetaat* (from *itek*, "toe piece of a boot")
ayuq/ayut	Labrador tea (from *ayu-*, "to spread, to go farther and farther away")

caiggluk/caiggluut	wormwood, stinkweed; also *qanganaruat* (lit., "imitation squirrels"), *naunrallraat* (lit., "bad plants")
caqlak/caqliit	roseroot/s; also *megtat neqait* (lit., "bumble-bee food"), *evegtat neqait* (lit., "bumblebee food")
ciilqaaq/ciilqaaret	fireweed
cuukvaguaq/cuukvaguat	alder/s (lit., "pretend *cuukvak* [pike]"); also *auguqsulit* (possibly from *auk*, "blood")
cuyakeggliq/cuyakegglit	diamond leaf willow/s (lit., "ones with good leaves"); also *cuyaqsuut* (K) (lit., "ones with good leaves"), *uqvigayagaat* (HBC) (lit., "small *ukviaret* [willows]")
cuyangaaraq/cuyangaaraat	willow leaf bud/s (lit., "early leaves")
cuyaq/cuyat	tobacco (lit., "leaf")
elnguq/elngut	birch tree/s (lit., "pliant wood," from *elngur-*, "to be thick but pliable")
enrilnguaq/enrilnguat	young, edible willow shoot/s (from *eneq*, "bone," lit., "ones similar to ones without bones")
ikiituk/ikiituut	wild celery
kangipluk/kangipluut	charcoal
kavqsuk/kaviqsuut	red-barked or little tree willow/s (from *kavir-*, "to be red")
kevraartuq/kevraartut	spruce tree/s; also *nekevraartut*
melquruaq/melquruat	white or Russet cottongrass (lit., "pretend *melquq* [fur]"); also *maqaruaruat* (lit., "pretend *maqaruat* [snowshoe hares]"), *ukaviruat* or *ukayiruat* (lit., "pretend *ukayiq* [cognate to Siberian Yupik *ukaziq*, snow-shoe hare]"), *qikmiruat*
pellukutaq/pellukutat	leaf/leaves of coltsfoot; also *qaltaruat* (lit., "pretend buckets"), *arakat* (lit., "ones that will be used for *araq* [ash]")
tayarulunguaq/tayarulunguat	horsetail plant/s (lit., "fake mare's tail")
teptukuyak/teptukuyiit	valerian (from *teptu-*, "to be odoriferous")
tumagliq/tumaglit	low bush cranberry/ies (from *tumag-* "to be bitter tasting"); also *kavirlit* (from *kavir-* "to be red," lit., "red ones")
uqvigpik/uqvigpiit	Alaska bog willow/s (lit., "real *ukvik* [willow]"); also *uqvigpiaq/uqvigpiat*
uqvik/uqviit	willow/s, *Salix* species; also *uqviaq/uqviaret*, *uqvigaq/uqvigaat*, *uqviaraq/uqviarat*

TUQUNARQELLRIIT NAUCETAAT
POISONOUS PLANTS

. . . .

anguturluq/anguteurluq/anguteurluut   male water hemlock (lit., "poor dear
male"); also *uquutvaguat*

pupignaq/pupignat   poisonous mushroom/s (lit., "one that
causes sores")

teggneurluut   plant part of poison water hemlock (lit.,
"poor dear *teggneret* [elders]")

tulukaruut neqait   baneberries (lit., "raven's food")

ATSAT
BERRIES

. . . .

agautaq/agautat   northern red currant/s (lit., "hanging
things")

atsaanglluk/atsaanglluut   northern black currant/s; also *atsangayiit*
(from *atsat*, "berries")

atsat/atsalugpiat   cloudberries, salmonberries (lit., "genuine
atsat [berries]"); also *naunrat*

cingqullektaq/cingqullektaat bunchberry/ies, air berry/ies (from *cingqur-*,
"to make loud popping noises")

curaq/curat   blueberry/ies

kavlak/kavliit   bearberry/ies; also *kavlagpiit*

kitngigpak/kitngigpiit   highbush cranberry/ies (lit., "big *kitngit*
[low bush cranberries]"); also *atsaangruyiit*
(from *atsat*, "berries"), *mercuullugpiit* (lit.,
"something big to drink water with")

paunraq/paunrat   blackberry/ies, crowberry/ies; also *tan'gerpiit* (lit., "black ones")

puyuraar/puyuraaraat   dwarf nagoonberry/ies; also *puyurnit*

tumagliq/tumaglit   low bush cranberry/ies; also *kavirlit* (lit.,
"red ones"), *kitngit* (Y)

uingiar/uingiaraat   bog cranberry/ies (lit., "husbandless
females"); also *quunarliaraat* (from
quunarqe-, "to taste sour"), *uskurtuliarat*
(NS) (lit., "ones with *uskurat* [tethers]")

CAN'GET
GRASSES

evepik/evepiit   bluejoint (lit., "real *evek* [grass]")

evisrayaaq/evisrayaaret   short grass/es (lit., "small *evek* [grass]"); also *evesraaraat*

kelugkaq/kelugkat   water sedge (from *keluk*, "stitch," lit., "those to be used as stitches")

qayikvayak/qayikvayiit   wideleaf polargrass/es

taperrnaq/taperrnat   rye grass/es, coarse seashore grass/es

UUGNARAAT NEQAUTAIT
MOUSE FOODS

aatuuqerrayak/aatuuqerrayiit   unidentified tubers collected by voles, said to look like small green Christmas trees; also *aatuuyaarpak* (NI)

agivaat/agivaat naunrat   unidentified tubers collected by voles

anlleq/anlleret   tall cottongrass tuber/s

negaasek/negaasget   tuber/s of Eged's silverweed; also *marallaat* (HBC) (lit., "ones from the *maraq* [marshy lowland]")

qetek/qet'get, qetgeret   root nodule/s of *tayarulunguat* (horsetail plants, lit., "fake mare's tail")

utngungssaq/utngungssaraat   grass seed/s, said by elders to be from *taperrnat* (rye grass) (from *utnguk*, "wart")

English and Latin Plant Names

Note: This list includes the currently accepted name for each species, generally employing those listed in *Flora of North America North of Mexico* (Flora of North America Editorial Committee, eds. 1993+) with additional information from *Flora of Alaska* (Ickert-Bond et al. 2019). In cases where there are commonly known synonyms, these are listed following the entry.

Common name(s)	Yup'ik name(s)	Latin name(s)
alders	*cuukvaguat, auguqsulit*	*Alnus viridis* subsp. *sinuata* synonym: *Alnus sinuata* *Alnus incana* subsp. *tenuifolia*
alpine sweet vetch, "Eskimo potato"	*elagat, qerqat, marallaat*	*Hedysarum alpinum* synonym: *Hedysarum americanum* (FAK)
aquatic plant, unidentified	*qecigpiit, qeciguut, qecigkat, qecigtut*	
baneberries	*tulukaruut neqait*	*Actaea rubra*
beach greens	*it'garalget, itegarat, qel-quayat, tukulleggat*	*Honckenya peploides*
bearberries	*kavliit, kavlagpiit*	*Arctous alpina* synonym: *Arctostaphylos alpina*
birch bracket fungus, punk	*arakat, kumakat*	*Fomitopsis betulina*
birch trees	*elngut*	*Betula papyrifera*
blackberries, crowberries	*paunrat, tan'gerpiit*	*Empetrum nigrum*
bluebells	*panayulit neqait kulukuunaruat*	*Mertensia paniculata*

Common name(s)	Yup'ik name(s)	Latin name(s)
blueberries	*curat*	*Vaccinium uliginosum*
bunchberries, air berries	*cingqullektaat*	*Cornus suecica* (over most of the region) *synonym: Chamaepericlymenum suecicum* (FAK) *Cornus canadensis* (upper Yukon and upper Kuskokwim) *synonym: Chamaepericlymenum canadensis* (FAK)
buttercups	*kapuukaraat, qaqacuqunat*	*Ranunculus pallasii* *synonym: Coptidium palasii* (FAK)
cloudberries, salmonberries	*atsat, atsalugpiat, naunrat*	*Rubus chamaemorus*
coltsfoot leaves	*pellukutat, qaltaruat, arakat*	*Petasites frigidus*
cottongrass, tall	*iitaat*	*Eriophorum angustifolium*
cottongrass, white or Russet	*melquruat, maqaruaruat, ukaviruat* or *ukayiruat, qikmiruat*	*Eriophorum scheuchzeri* *Eriophorum chamissonis* synonym: *Eriophorum russeolum*
cow parsnips	*tarnat*	*Heracleum lanatum*
cranberries, bog	*uingiaraat, quunarliaraat, uskurtuliarat*	*Vaccinium oxycoccos* synonym: *Oxycoccus microcarpus* (FAK)
cranberries, highbush	*kitngigpiit, atsaangruyiit, mercuullugpiit*	*Viburnum edule*
cranberries, low bush	*tumaglit, kavirlit, kitngit*	*Vaccinium vitis-idaea*
dwarf nagoonberries	*puyuraat, puyuraaraat, puyurnit*	*Rubus arcticus*
ferns, fiddlehead and wood ferns	*cetuguat, cetugpaguat, ceturqaaraat*	*Athyrium filix-femina* *Dryopteris expansa* synonym: *Dryopteris dilatata*
fireweed	*ciilqaaret*	*Chamerion angustifolium* synonym: *Epilobium angustifolium*
grass, bluejoint	*evepiit*	*Calamagrostis canadensis*
grass, rye; coarse seashore grass	*taperrnat*	*Leymus mollis* synonyms: *Elymus mollis, Elymus arenarius subsp. mollis*
grass seeds	*utngungssaraat*	*Poaceae* [unidentified]
horsetail plants	*tayarulunguat*	*Equisetum* spp.
Labrador tea	*ayut*	*Rhododendron tomentosum* synonym: *Ledum palustre*
lichens, kidney	*ciruneruat*	*Nephroma* spp.
lichens, reindeer	*taqukanguat, tuntut neqait, ungagat, qilungayagaat*	*Cladonia rangiferina*

Common name(s)	Yup'ik name(s)	Latin name(s)
mare's tail	*tayarut, tayarum iluraa, tayarulungniit*	*Hippuris tetraphylla* (= *tayarut*) *Hippuris vulgaris* (= *tayarum iluraa*)
marsh fivefinger plants	*pingayunelget*	*Comarum palustre* synonym: *Potentilla palustris*
marsh marigolds	*allngiguat, allmaguat, irunguat, uivlut*	*Caltha palustris*
mosses, sphagnum	*urut*	*Sphagnum* spp.
mountain sorrel	*quunarliaraat*	*Oxyria digyna*
mushrooms, edible	*palurutat* [general category]	
mushrooms, poisonous	*pupignat* [general category]	
northern black currants	*atsaanglluut, atsangayiit*	*Ribes hudsonianum*
northern red currants	*agautat*	*Ribes triste*
nutty saw-wort	*nasqupaguat*	*Saussurea nuda*
oysterleafs, beach bluebells	*neqnirliaraat*	*Mertensia maritima*
poison water hemlock, male	*anguteurluut, uquutvaguat*	*Cicuta virosa* synonym: *Cicuta mackenzieana*
puffball fungus	*agyat anait*	*Lycoperdon* spp.
rockweed	*elquat epuit*	*Fucus gardneri*
root nodules, horsetail	*qet'get, qetgeret*	*Equisetum* spp.
rose hips	*tuutaruat*	*Rosa acicularis*
roseroot	*caqliit, megtat neqait, evegtat neqait*	*Rhodiola integrifolia* synonym: *Sedum rosea* subsp. *integrifolium*
saxifrage	*mecurtuliaraat, mecugtuli-araat*	*Micranthes* spp., *Saxifraga* spp.
sea lovage	*mecuqelugaat, mercurtulgaat, mecurtulit, mercurtuliaraat, mecuqelluqaat, mecugglug-gaat*	*Ligusticum scoticum*
sour dock	*quagcit, aatunat, quunarlit*	*Rumex arcticus*
spreading wood fern roots	*kun'at*	*Dryopteris expansa* synonym: *Dryopteris dilatata* (FAK)
spruce trees	*kevraartut, nekevraartut*	*Picea glauca*
stinging nettles	*qatlinat, qacelpiit, qacellinat, mingqutnguayaalget*	*Urtica dioica* subsp. *gracilis* synonym: *Urtica gracilis*
tobacco	*cuyat*	*Nicotiana tabacum*
tuberous spring beauty, Eskimo potatoes	*ulqit, atkallaat*	*Claytonia tuberosa*

Common name(s)	Yup'ik name(s)	Latin name(s)
tubers, tall cotton grass	anlleret	Eriophorum angustifolium
tubers, silverweed	negaasget, marallaat	Potentilla anserina subsp. groenlandica synonyms: Argentina anserina subsp. egedei, Potentilla egedei
tubers, unidentified	aatuuqerrayiit	N/A
tubers, unidentified	agivaat, agiyaat naunrat	N/A
valerian	teptukuyiit	Valeriana capitata
water sedge	kelugkat	Carex spp.
water weed roots water weeds	muugarliarneret, nuyaruat	Sparganium angustifolium Sparganium hyperboreum
wideleaf polargrass	qayikvayiit	Arctagrostis latifolia
wild celery	ikiituut	Angelica lucida synonym: Angelica gmelinii (FAK)
wild chamomile, pineapple weed	atsaruat, itemkeciyaat, kitengkaciyaaret, iteg-migcetaat	Matricaria discoidea synonym: Matricaria matricarioides
wild rhubarb	angukat, nakaaret, arnaurluut	Aconogonon alaskanum synonym: Polygonum alaskanum
willow leaf buds	cuyangaaraat	Salix spp.
willow shoots, young	enrilnguat	Salix spp.
willows	uqvigaat, uqviaret	Salix spp.
willows, Alaska bog	uqvigpiit	Salix fuscescens
willows, diamond leaf	cuyakegglit, cuyaqsuut, uqvi-gayagaat	Salix pulchra
willows, red-barked or little tree	kaviqsuut	Salix arbusculoides
wood ash	araq, qamlleq	N/A
woolly lousewort	ulevleruyiit	Pedicularis spp.
wormwood, stinkweed	caiggluut, qanganaruat, naunrallraat	Artemisia tilesii
yarrow	anuqetuliaraat, anguqe-tussngit	Achillea millefolium Achillea alpina (Siberian yarrow) synonym: Achillea sibirica

Notes

1 Audio recordings of these gatherings, as well as recordings I made with Yup'ik men
and women between 1976 and 2000, are archived at the Alaska Native Language
Archives, University of Alaska Fairbanks, along with an index of written tran-
scripts, organized by date, including where the recording was made, participants,
and translator. Full references for all passages quoted in the bilingual section of this
book are given as endnotes to give the reader a better picture of the many participants involved in each recording.

2 Technically, fungi are in their own kingdom, Fungi. True plants are in the kingdom
Plantae, algae are in the kingdom Protista, and lichens are a hybrid combination of
fungi and algae. All of these used to be considered "plants," but this is no longer the
case.

3 Even scientific Latin names can vary, and there is a complicated synonymy created
by different authorities using different names to describe the same species. These
changing names often result from modern genetics (Torre Jorgenson, personal
communication, May 2019).

4 The related compound thujone is more prevalent in other *Artemisia* species, but it
has a different effect and is more of a stimulant (Overfield et al. 1980:97).

5 July 2007:247; *CEC Nelson Island Circumnavigation* with Simeon and Anna Agnes,
Nightmute; Paul and Martina John and Joe Felix, Toksook Bay; Theresa Abraham,
Chefornak; Michael and John Roy John, Newtok; Rita Angaiak, John Walter Sr.,
and John Walter Jr. of Tununak; June McAtee, Calista; Steve Street, AVCP; Tom
Doolittle, USFWS; four students, village helpers, and CEC staff Ruth Jimmie, David
Chanar, Alice Rearden, and Ann Riordan; twenty-one tapes transcribed and trans-
lated by Alice Rearden.

6 November 5–6, 2007:321–24; *CEC Nelson Island Women's Gathering*, Whitehouse
B&B, Bethel. Albertina Dull, Nightmute; Martina John, Toksook Bay; Helen Walter,
Tununak; Theresa Abraham, Chefornak, with Alice Rearden and Ann Riordan; nine
tapes transcribed and translated by Alice Rearden.

7 January 5–6, 2011:239–40; *Quinhagak Elders Gathering*, Ann Riordan's house,
Anchorage, with Annie Cleveland, Martha Mark, Joshua Cleveland, George
Pleasant, Pauline Matthew, Warren Jones, and Ann Riordan; eight tapes tran-
scribed and translated by Alice Rearden.

8 January 5–6, 2011:242; *Quinhagak Elders Gathering*, Ann Riordan's house, Anchorage,
with Annie Cleveland, Martha Mark, Joshua Cleveland, George Pleasant, Pauline
Matthew, Warren Jones, and Ann Riordan; eight tapes transcribed and translated
by Alice Rearden.

9 November 5–6, 2007:312; *CEC Nelson Island Women's Gathering*, Whitehouse B&B,
Bethel, with Albertina Dull, Nightmute; Martina John, Toksook Bay; Helen Walter,
Tununak; Theresa Abraham, Chefornak, and Alice Rearden and Ann Riordan; nine
tapes transcribed and translated by Alice Rearden.

10 July 2007:239–40; *CEC Nelson Island Circumnavigation* with Simeon and Anna
Agnes, Nightmute; Paul and Martina John and Joe Felix, Toksook Bay; Theresa
Abraham, Chefornak; Michael and John Roy John, Newtok; Rita Angaiak, John
Walter Sr., and John Walter Jr. of Tununak; June McAtee, Calista; Steve Street,
AVCP; Tom Doolittle, USFWS; four students, village helpers, and CEC staff Ruth
Jimmie, David Chanar, Alice Rearden, and Ann Riordan; twenty-one tapes tran-
scribed and translated by Alice Rearden.

11 October 21–22, 2003:123; *CEC Survival Skills Gathering*, Anchorage Museum,
Anchorage, with Frank Andrew, Kwigillingok; Paul John, Toksook Bay; John Phillip,
Kongiganak; Peter Jacobs, Bethel; Nick Andrew, Marshall; Irvin Brink, Kasigluk,
with Alice Rearden and Ann Riordan; seven tapes transcribed and translated by
Alice Rearden.

12 January 24–26, 2007:589–93; *CEC Toksook Elders' Gathering*, Toksook Bay
Traditional Council Office, Toksook Bay. Lizzie Chimiugak, Sophie Agimuk, Martina
John, John Alirkar, Phillip Moses, with David Chanar and Ann Riordan; ten tapes
transcribed and translated by Alice Rearden and David Chanar.

13 David Chanar, originally from Toksook Bay and also taking part in the conversation, provided these literary translations for *ngelaq'erluni* (chuckling) by his uncle
Phillip, who was telling the story. A fine translator with a flair for the dramatic,

David invariably sought out colorful ways of conveying the fun of storytelling.

14 June 23, 2009:155–58; *CEC Interview with Albertina Dull, Nightmute, and her sister Lizzie Chimiugak, Toksook Bay*, with Ruth Jimmie and Ann Riordan at Lizzie's home in Toksook Bay; six tapes transcribed and translated by Alice Rearden.

15 April 10–12, 2012:61–66; *Lower Yukon Women's Gathering*, Anchorage, with Barbara Joe, Maryann Andrews, Mary and Peter Black, Mark John, Marie Meade, Alice Rearden, and Ann Riordan; twelve tapes transcribed and translated by Alice Rearden.

16 September 24–26, 2003:227–32; *CEC Materials Gathering*, Pacifica Guest House, Bethel. Paul John and Theresa Moses, Toksook Bay; Wassillie B. Evan, Akiak; Tim and Marie Myers, Pilot Station with KYUK staff Jackie Cleveland, Dean Swope, and Mike Martz and CEC staff Freda Jimmie, Denis Sheldon, and Ann Riordan; eight tapes transcribed and translated by Alice Rearden.

17 September 24–26, 2003:227–32; *CEC Materials Gathering*, Pacifica Guest House, Bethel. Paul John and Theresa Moses, Toksook Bay; Wassillie B. Evan, Akiak; Tim and Marie Myers, Pilot Station with KYUK staff Jackie Cleveland, Dean Swope, and Mike Martz and CEC staff Freda Jimmie, Denis Sheldon, and Ann Riordan; eight tapes translated and transcribed by Alice Rearden.

18 January 15–17, 2013:351–53; *Lower Yukon Men's Gathering*, Ann Riordan's house, Anchorage, with Raphael Jimmy, Denis Sheldon, Francis Charlie, Mark John, Alice Rearden, and Ann Riordan; eleven tapes transcribed and translated by Alice Rearden.

19 January 15–17, 2013:354–57; *Lower Yukon Men's Gathering*, Ann Riordan's house, Anchorage, with Raphael Jimmy, Denis Sheldon, Francis Charlie, Mark John, Alice Rearden, and Ann Riordan; eleven tapes transcribed and translated by Alice Rearden.

20 May 6–8, 2004:100; *CEC Harvesting Gathering #2*, Pacifica Guest House, Bethel, with Theresa Moses and Paul John, Toksook Bay; David Martin, Kipnuk; Neva Rivers, Hooper Bay, and CEC staff Mark John, Marie Meade, and Ann Riordan; eleven tapes transcribed and translated by Alice Rearden.

21 March 31–April 1, 2004:556–59; *CEC Harvesting Gathering*, Pacifica Guest House, Bethel; Nick Andrew, Marshall; Wassillie B. Evan, Akiak; Paul John, Toksook Bay; Benedict Tucker, Emmonak; Neva Rivers, Hooper Bay, with Marie Meade and Ann Riordan. twelve tapes transcribed and translated by Alice Rearden.

22 April 30–May 1, 2019:289; *Bethel Plant Gathering*, Bethel, AK, USFWS Bunkhouse, with John Andrew, Cecelia Andrews, Theresa Abraham, Ruth Jimmie, Mark John, Corey Joseph, and Ann Riordan; seven tapes transcribed and translated by Alice Rearden.

23 March 4, 2007:72; *Chefornak Elders' Gathering*, Chefornak Community Hall; John Eric, Paul Tunuchuk, David and Pauline Jimmie, Maria Eric, and Martina Wasili, with Mark John, David Chanar, and Ann Riordan; ten tapes transcribed and translated by Alice Rearden.

24 October 16–18, 2010:125–27, *National Park Serviec Regional Gathering*, Ann Riordan's house, Anchorage, with Paul John, John Phillip, Nick Andrew, Martin Moore, Bob Aloysius, Moses Paukan, Mark John, Alice Rearden, and Ann Riordan; ten tapes transcribed and translated by Alice Rearden.

25 May 6–8, 2004:98–99; *CEC Harvesting Gathering #2*, Pacifica Guest House, Bethel; Theresa Moses and Paul John, Toksook Bay; David Martin, Kipnuk; Neva Rivers, Hooper Bay, with CEC staff Mark John, Marie Meade, and Ann Riordan; eleven tapes transcribed and translated by Alice Rearden.

26 April 30–May 1, 2019:175–79; *Bethel Plant Gathering*, Bethel, AK, USFWS Bunkhouse, with John Andrew, Cecelia Andrews, Theresa Abraham, Ruth Jimmie, Mark John, Corey Joseph, and Ann Riordan; seven tapes transcribed and translated by Alice Rearden.

27 July 2007:239–40; *CEC Nelson Island Circumnavigation*, with Simeon and Anna Agnes, Nightmute; Paul and Martina John and Joe Felix, Toksook Bay; Theresa Abraham, Chefornak; Michael and John Roy John, Newtok; Rita Angaiak, John Walter Sr., and John Walter Jr., Tununak; June McAtee, Calista; Steve Street, AVCP; Tom Doolittle, USFWS; four students, village helpers, and CEC staff Ruth Jimmie, David Chanar, Alice Rearden, and Ann Riordan; twenty-one tapes transcribed and translated by Alice Rearden.

28 October 6–7, 2009; *Quinhagak Elders Gathering*, Quinhagak, with Annie Cleveland, Paul Beebe, Florence Jones, Nick Mark, Carrie Pleasant, Alice Mark, Martha Mark, Willie Mark, Joshua Cleveland, George Pleasant, Pauline Matthew, Warren Jones, and Ann Riordan; seven tapes transcribed and translated by Alice Rearden.

29 June 3-4, 2016: 253. *Akulmiut Women's Gathering,* Nunapitchuk, with Katie Jenkins, Marie Alexie, Grace Parks, Marie Meade, and Ann Riordan; seven tapes transcribed and translated by Alice Rearden.

References

Ager, Thomas, and Lynn Price Ager. 1980. "Ethnobotany of the Eskimos of Nelson Island, Alaska." *Arctic Anthropology* 17:27–48.

Ainana, Lyudmila, and Igor Zagrebin. 2014. *Edible Plants Used by Siberian Yupik Eskimos of Southeastern Chukchi Peninsula, Russia*. English translation by Richard L. Bland. Anchorage: National Park Service, Shared Beringian Heritage Program.

Becker, A. L. 2000. *Beyond Translation: Essays toward a Modern Philology*. Ann Arbor: University of Michigan Press.

Fienup-Riordan, Ann. 1994. *Boundaries and Passages: Rule and Ritual in Yup'ik Eskimo Oral Tradition*. Norman: University of Oklahoma Press.

———. 1996. *The Living Tradition of Yup'ik Masks: Agayuliyararput/Our Way of Making Prayer*. Seattle: University of Washington Press.

———. 2007. *Yuungnaqpiallerput/The Way We Genuinely Live: Masterworks of Yup'ik Science and Survival*. Seattle: University of Washington Press.

Fienup-Riordan, Ann, and Alice Rearden. 2017. *Qanemcit Amllertut/Many Stories to Tell: Tales of Humans and Animals from Southwest Alaska*. Fairbanks: University of Alaska Press and the Alaska Native Language Center.

Flora of North America Editorial Committee, eds. 1993+. *Flora of North America North of Mexico*. 19+ volumes. New York and Oxford.

Garibaldi, Ann. 1999. *Medicinal Flora of the Alaska Natives*. Alaska Natural Heritage Program, Environment and Natural Resources Institute, University of Alaska, Anchorage.

Gray, Beverly. 2011. *The Boreal Herbal: Wild Food and Medicine Plants of the North*. Whitehorse, Yukon, Canada: Aroma Borealis Press.

Griffin, Dennis. 2001. "Contributions to the Ethnobotany of the Cup'it Eskimo, Nunivak Island, Alaska." *Journal of Ethnobiology* 21 (2): 91–127.

———. 2008. "Ethnobotany of Alaska: A Southwestern Alaska Perspective." In *Encyclopedia of History of Science, Technology, and Medicine in Non-Western Cultures*, 813–26. Berlin: Springer-Verlag.

———. 2009. "The Ethnobiology of the Central Yup'ik Eskimo, Southwestern Alaska." *Alaska Journal of Anthropology* 7 (2): 81–100.

Hymes, Dell. 1981. *"In Vain I Tried to Tell You": Essays in Native American Ethnopoetics.* Studies in Native American Literature 1. Philadelphia: University of Pennsylvania Press.

Ickert-Bond, S.M., B. Bennett, M.L. Carlson, J. DeLapp, J.R. Fulkerson, C.L. Parker, T.W. Nawrocki, M.C. Stensvold, and C.O. Webb (eds.). 2019. *Flora of Alaska.* Available https://floraofalaska.org.

Jacobson, Steven A. 1984. *Yup'ik Eskimo Dictionary.* Fairbanks: Alaska Native Language Center, University of Alaska.

———. 1995. *A Practical Grammar of the Central Alaskan Yup'ik Eskimo Language.* Fairbanks: Alaska Native Language Center, University of Alaska.

———. 2012. *Yup'ik Eskimo Dictionary.* 2nd ed. Fairbanks: Alaska Native Language Center, University of Alaska.

Jernigan, Kevin, with Oscar Alexie, Sophie Alexie, Michelle Stover, Rose Meier, Carolyn Parker, Mary Pete, Memmi Rasmussen, Rose Domnick, and Alice Fredson, eds. 2015. "A Guide to the Ethnobotany of the Yukon-Kuskokwim Region." Alaska Native Language Center online manual.

Jones, Anore. 2010. *Plants That We Eat/Nauriat Nigiñaqtuat: From the Traditional Wisdom of the Iñupiat Elders of Northwest Alaska.* Fairbanks: University of Alaska Press.

Lantis, Margaret. 1959. "Folk Medicine and Hygiene: Lower Kuskokwim and Nunivak-Nelson Island Areas." *Anthropological Papers of the University of Alaska.* 8 (1): 1–76.

Mather, Elsie P. 1995. "With a Vision beyond Our Immediate Needs: Oral Traditions in an Age of Literacy." In *When Our Words Return: Writing, Hearing, and Remembering Oral Traditions of Alaska and the Yukon.* Edited by Phyllis Morrow and William Schneider, pp. 13–26. Logan: Utah State University Press.

Miyaoka, Osahito, and Elsie Mather. 1979. *Yup'ik Eskimo Orthography.* Bethel, AK: Kuskokwim Community College.

Nelson, Edward William. 1899. *The Eskimo about Bering Strait.* Bureau of American Ethnology Annual Report for 1896–1897. Vol. 18, Pt. I. Washington, DC: Smithsonian Institution Press (Reprinted 1983).

Nuniwarmiut Taqnelluit (Elders of Nunivak Island). 2018. *Nuniwami Navcit Cenallat-llu/Nunivak Plant and Seashore Life: The Ethnobotany of the Nuniwarmiut Eskimo, Nunivak Island, Alaska.* Compiled and edited by Dennis Griffin. Fairbanks: Alaska Native Language Center, University of Alaska.

Oswalt, Wendell. 1957. "A Western Eskimo Ethnobotany." *Anthropological Papers of the University of Alaska.* 6 (1): 17–36.

Overfield, T., W. W. Epstein, and L. A. Gaudioso. 1980. "Eskimo Uses of *Artemisia tilesii* (Compositae)." *Economic Botany* 34 (2): 97–100.

Reed, Irene, Osahito Miyaoka, Steven Jacobson, Pascal Afcan, and Michael Krauss. 1977. *Yup'ik Eskimo Grammar.* Fairbanks: Alaska Native Language Center, University of Alaska.

Robuck, O. Wayne. 1985. *The Common Plants of the Muskegs of Southeast Alaska*. US Department of the Interior, Forest Service Pacific Northwest Forest and Range Experiment Station.

Russell, Priscilla N. 2017. *Naut'staarpet/Our Plants: A Kodiak Alutiiq Plantlore*. Kodiak, AK: Alutiiq Museum and Archaeological Repository.

Schofield, Janice J. 2011. *Alaska Wild Plants: A Guide to Alaska's Edible Harvest*. Portland, OR: Alaska Northwest Books.

———. 1989. *Discovering Wild Plants: Alaska, Western Canada, the Northwest*. Anchorage: Alaska Northwest Books.

Swann, Brian. 1994. *Coming to Light: Contemporary Translations of the Native Literatures of North America*. New York: Random House.

Tedlock, Dennis. 1983. *The Spoken Word and the Work of Interpretation*. Philadelphia: University of Pennsylvania Press.

Unger, Suanne, with Mary Bourdukofsky, Moses Dirks, Sally Swetzof, and Douglas Veltre. 2014. *QaqamiiĝuX̂: Traditional Foods and Recipes from the Aleutian and Pribilof Islands*. Anchorage: Aleutian Pribilof Islands Association, Inc.

Woodbury, Anthony C. 1984. "Eskimo and Aleut Languages." In *Arctic*, Vol. 5, *Handbook of North American Indians*. Edited by David Damas, 49–63. Washington, DC: Smithsonian Institution Press.

The upper Kwethluk River in fall, 2007, where John Andrew traveled and hunted when he was young. JOSH SPICE, US FISH AND WILDLIFE SERVICE

Index

Note: page numbers in italics refer to figures;
those followed by *t* or m refer to tables and maps respectively.

cooking and eating of, 32, 33–34,
216–17
as edible plant, 296
etymology of Yup'ik names for, 29,
34, 296
gathering of, 24, 32, 76, 214–17,
242–43
mature plant, 30
second edible growth, 32
time for gathering, 27, 28–31, 33,
214–17
tool used to harvest, 24, 29–31, 32,
214–17
Yup'ik names for, 296, 302t

caagnitellriit (people in life transitions)
eyagyarat (abstinence practices) by,
20, 156
spruce trees as supernatural danger
to, 156
caiggluk/caiggluut. See wormwood [stink-
weed]
Cakcaaq River, 207
Calamagrostis canadensis. See grass,
bluejoint
Calista Education and Culture. *See* Calista
Elders Council
Calista Elders Council (CEC)
documentation of history and oral
traditions, 3
gatherings and field trips by, 3
Cal'itmiut
gathering from vole caches in, 202
tradition bearers born in, xx
Caltha palustris. See marsh marigolds
can'get. See grass(es)
Caniliaq, tradition bearers born in, xix
Canineq area
CEC documenting of history and
culture in, 3
plants found in, 22
Yup'ik plant names in, 295–300
See also Kuskokwim River, lower
caqlak/caqliit. See roseroot
Carex spp. *See* water sedge
caribou
blood, as cooking ingredient, 80
dishes using, 274–75
plants eaten by, 62, 91-92
Carl, Susie, *188, 192*, 272–73

Carter, Joe, 127
caulk for kayaks, 89, 90, 230–31
CEC. *See* Calista Elders Council
cetugpaguat. See ferns, fiddlehead and
wood ferns
cetuguaq/cetuguat. See ferns, fiddlehead
and wood ferns
ceturqaaraat. See ferns, fiddlehead and
wood ferns
Ceturrnaq River, gathering from vole
caches along, 191
Cevv'arneq, 5m
berry picking near, 180
tradition bearers born in, xx
Chamaepericlymenum canadensis. See
bunchberries [air berries]
Chamaepericlymenum suecicum. See bunchberries [air berries]
Chamerion angustifolium. See fireweed
Chanar, Brentina (*Papangluar*), xx, 134
Chanar, David (*Cingurruk*), xxi, 208,
230–31
charcoal (*kangipluk*)
carrying of, as protection, 158
as compress, 158
Charles, Nick, Sr. (*Ayagina'ar*), xxi, 27
Charlie, Francis (*Acqaq*), xix
on alpine sweet vetch, eating of, 54
on ash, medicinal properties of,
157–58
on coltsfoot, 141–42
on fiddlehead ferns, cooking with, 46
on gathering of plants, 102–3
on horsetail root nodules, 199
on Labrador tea, 117, 118, 254–63
on tall cottongrass tubers, 194
on water weed roots, eating of, 56–57
on wild celery, 136
on willow as medicinal plant, 149
on wormwood as medicinal plant, 109
Chefornak, 5m
berry picking near, *284–85*
Maria Erik, 28
Pauline Jimmie, 79
picking of salmonberries near, 172
plants gathered in spring, 28
Theresa Abraham, 18
and trade in ash from burnt wood, 157
tradition bearers residing in, xx
wormwood as medicinal plant in, 104

ircenrraat (other-than-human persons) (*cont.*)
  three doorways of, 266–67
  voles as, 194, 280–81
irunguat. See marsh marigolds
issran (grass carrying bag), 173, 175
itegarat. See beach greens
itegmigcetaat. See wild chamomile [pineapple weed]
itemkeciyaat. See wild chamomile [pineapple weed]
it'garalek/it'garalget. See beach greens
ivrucik (skin boots), 42

Jackson, Annie, xxi, 89, 127, 146
Jacob, Fannie (*Mayuralria*), xxi, 142
Jacobs, Peter, Sr. (*Paniguaq*), xxi
  on factors affecting berry harvest, 169
  on salmonberries, time for picking of, 170
  on wild chamomile, 123
  on wild rhubarb and sour dock as mates, 72, 75, 226–27
  on wild rhubarb height, 75, 226–27
Jenkins, Katie (*Uruvak*), xxi, 126, 177, 290–91
Jernigan, Kevin, 2, 4–7, 83
  on bumblebee food, 65
  on cow parsnips, 133
  on fiddlehead ferns, names for, 46
  on names for sour dock, 65
  photos by, 41, 44, 45, 49, 52, 56, 59, 60, 62, 64, 71, 73, 76, 78, 82, 84, 87, 92, 101, 102, 111, 119, 124, 127, 128, 131, 132, 133, 134, 138, 139, 141, 143, 145, 148, 153, 155, 159, 163, 166–67, 173, 178, 181, 182, 184, 185, 187, 195, 200
  on puffball mushrooms, antimicrobial powers of, 159
  on rockweed, 52
  on sea lovage, Yup'ik terms for, 49
  on sour dock, medicinal uses of, 71
  on valerian, 142
  on yarrow, 137–39
Jimmie, Pauline (*Kangrilnguq*), xx
  on grass seeds, cooking with, 196
  on mare's tail as emergency food in famine, 79–80, 224–25
  on mouse foods, 274–75
  on silverweed tubers, cooking with, 196

Jimmie, Ruth (*Angalgaq*), xx
  on bearberries, 183–84
  on berry picking, 177
  on bluebells, eating of, 65
  on eating of pussy willows, 147
  on fireweed, 129
  and gathering of plants, 12–13, 37, 38
  on grass seeds, cooking with, 196
  on legend about origin of poison water hemlock, 238–39
  on lichens as emergency food in famine, 92
  on low bush cranberries, 125
  on mushrooms, eating of, 93
  photo by, 8–9
  on sea lovage, eating of, 51
  on voles, 194, 278–83
  on wild chamomile, 123
  on willow tips, chewing of, 147
  on wormwood as medicinal plant, 109–10
Jimmy, Raphael (*Angagaq*), xxi
  on alpine sweet vetch, eating of, 54
  on cow parsnips, 133
  on gathering of plants, 54
  on Labrador tea, 111, 116, 118, 256–63
  on male and female parts of plants, 22–23
  on tall cottongrass tubers, 194
  on voles, 189–90
  on willows, medicinal properties of, 148–49
Jimmy, Ruth
  on bearberries, cooking with, 183
  on coltsfoot leaf buckets, 142
  on gathering of plants, 46
Joe, Barbara (*Arnaucuaq*), xix
  on Labrador tea, 111, 115
  on medicinal plants, 242–49
  on plants as food and medicine, 100
  on pure people, sensitivity to traditional medicine, 246–47
John, Mark (*Miisaq*), xxi
  on mare's tail, cooking with, 81
  on rockweed, eating of, 52
John, Martina (*Anguyaluk*), xx
  on berry picking, 175
  on gathering of plants, 38, 76, 216–17
  on low bush cranberries, 125
  on mountain sorrel, taste of, 72

male (*angucaluut*) and female
(*arnacaluut*) parts of, 23, 102–3
as medicinal plant, 248–51, 298
bathing in smoke for ritual cleansing (*tarvaq*), 110
belief as necessary for effect,
248–49, 250–51
chewing of, 106, 250–51
codeine-like properties of isothujone in, 104
for cuts, boils, and sores, 107, 126,
248–49
increased power with plant growth,
106
inhalation via steam, 107–8, 248–49
male *vs.* female parts and, 102–3
parts of plant used in, 104–5
as poultice, 109
preparation as tea, 104–6
to prevent colds, 106
for promotion of general health,
109–10
purgative sweating caused by, 107–8
for sore joints, 156–57, 248–49
with spruce, as tonic, 154
in steam baths, 108–9, 248–51
tea made from, 248–49
variety of uses, 104
as widely used, 101
smell of, 108, 250–51
tea made from, 248–49
time for gathering of, 103–4
treating of fish nets with, 142
various species globally, 104
Yup'ik names for, 21, 101–2, 248–49,
298, 304*t*

Yaayuk. See Hunt, Angela
yarrow (*Achillea millefolium*; *Achillea
alpina* [Siberian yarrow]), 137, 137–39,
138, 304*t*
etymology of Yup'ik names for, 297
as medicinal plant, 137–39, 297
millefolium vs. alpina [Siberian yar-
row] species, 137
tea made from, 139
Yup'ik names for, 137, 297, 304*t*
Yukon, Yup'ik plant names in, 295–300
Yukon-Kuskokwim region, 4m
studies on plant use in, 2

Yukon River
lower
CEC documenting of history and
culture in, 3
and Central Alaska Yup'ik language,
206
name for alpine sweet vetch, 54
plants found along, 53
shoreline of, 13
mouth of
gathering of berries, 6–7
storage of berries, 176
name for kidney lichens along, 91
plants, 22
between Russian Mission and
Mountain Village, 3
trees, 13
use of medicinal ash from tree fungus, 157
variations in plant names, 21
Yup'ik food, healthiness of, 99
Yup'ik knowledge
dwindling of, 21
as extensive, 7
of plants, variation by village, 22
Yup'ik language of Central Alaska
areas spoken, 206
demonstratives in, 210
five dialects of, 206
limited gender designations in, 209
nonspecific pronouns and phrases in,
209–10
as one of four Yupik languages, 206
orthography, 206–7
and place names, –*miut* ending on,
210
repetition, rhetorical function of,
208
issues in, 208–10
translators for this book, 208, 210,
211
verb tenses in, 209
word order in, 208–9
Yupik languages
Central Alaskan Yup'ik language as
one of, 206
and Eskimo-Aleut family of languages, 206
relation to Inuit/Iñupiaq languages,
206

Yup'ik names for plants
    great variation in, 2, 21
    information contained in, 22
    plants without English equivalents,
      21

Yup'ik names for plant parts, 22
Yuungnaqpiallerput/The Way We Genuinely Live (2007 exhibition), 3–4

Zagrebin, Igor, 1

CPSIA information can be obtained
at www.ICGtesting.com
Printed in the USA
LVHW070614100421
684029LV00011B/73

Under the Spell of
SUCCULENTS

JEFF MOORE

Dudleya brittoni with *Agave shawii* in northern Baja, Mexico. Photo by Jeremy Spath.

TO MY LOVELY WIFE LORI.

Whose work and dedication to our family has allowed me to have a job I love going to each day, despite the limited financial upside. Thank you sweetheart.

Under the Spell of Succulents: A Sampler of the Diversity of Succulents in Cultivation / Jeff Moore (3rd printing)

Text and photographs copyright © Jeff Moore 2016
All other photographers credited individually
Plant illustrations copyright © Irina Gronborg
Woodblock print on title page © Evelyn La Rosa

Jeff Moore, Solana Succulents
solanasucculents@sbcglobal.net
www.solanasucculents.com

ISBN: 978-0-9915846-0-4

Book design: Adrienne Joy

Printed and bound in Malaysia by Tien Wah Press

ABOUT THE PHOTOGRAPHY

Unless individually credited, most of the photographs in this book were taken by the author—an amateur photographer at best. I have been taking succulent photographs for years, as a hobby and also for garden club presentations. Part of the impetus for writing this book was to put all my images to good use. In my favor, more than any technical skill, is just an eye for a good shot. Most importantly, I have the advantage of always being around succulents, either at my nursery or during visits to a client's home or a grower's yard, or simply gardens in my area.

I made the decision to forgo the high-end DSLR and the bag full of lenses and tripod in favor of a high-end point-and-shoot. That may be an insult to my camera, a Cannon G10 that is capable of a lot more than I ask of it. I set it mostly on Auto, am aware of light conditions and backgrounds, find the plants I'm looking for (or whatever grabs my attention), and fire off a bunch of shots. Using digital, you can just keep going until you capture an image you like. For those of us used to waiting and paying for a pack of developed film at the drugstore, this is too good to be true. If you are a photography enthusiast and also like succulents, they are there just waiting for you—wonderful still-life subjects that sometimes defy the imagination. Get in close with a macro setting.

Some of the undersea succulent photos and cactus flower images were taken by Bob Wigand. He has the camera bag with all the goodies and requisite know-how. It shows.

ACKNOWLEDGMENTS

My list of individuals to thank is too long for this brief page, so I've expanded my thank-you list to the back. But my primary contributors in bringing this book to life are:

Adrienne Joy Armstrong, who took on the task of re-designing my initial effort into what you are looking at now. Her design took what was a sensory-overload photo-blast and turned it into a more refined finished product. Not being a succulent enthusiast, she was able to see the material through beginner's eyes and occasionally over-ride my enthusiasm for something obscure, in favor of something more appealing.

Ron Regehr of Cactus Canyon Nursery was my first proof-reader with plant experience. He helped with plant identification and certain grammatical issues, along with nomenclature. He really let loose with a red pen on a printed copy. I though he might find a couple of mistakes. He did, on the first page. I learned you can't proof your own work.

A special thanks to Debra Lee Baldwin, who like Ron is a succulent enthusiast, as evidenced by her three best-selling succulent books from Timber Press. She graciously included a nice foreword to this book along with some helpful advice regarding nomenclature and the challenges of putting a professional book into print. I've known Debra Lee for some time, and helped her a bit with her first books. Check out her website: www.debraleebaldwin.com. This is a self-published book, so any errors or omissions are strictly on the author.

In fact, the title of this book came from a discussion with writer Peter Jensen about Debra Lee. I mentioned to Peter that in researching her first book she made a few visits to my nursery, and it was quite apparent that she had 'come under the spell of succulents'. Peter immediately said "there's your book title". He was at least the third person that seemed to think I was going to write a book, so I took it as a sign that maybe I should write a book. So I did, and here it is. I hope you like it.

And lastly, Irina Gronborg, for her wonderful illustrations.

FACING PAGE *Agave* 'Red Margin', golden barrel cactus (*Echinocactus grusonii*) and black *Aeonium* 'Zwartkop'.

Contents

FOREWORD

Jeff Moore's plant nursery, Solana Succulents—located in the north San Diego community of Solana Beach—has been the go-to place for odd and unusual succulents for two decades. When you go there and have the eerie feeling you're being watched, it's likely by a pet reptile basking in the orange glow of a terrarium's heat lamp.

Yet of all the "plant guys" I've known, Jeff is arguably the least eccentric. He's devoted to his family, succulents and ocean sports. When I was a freelance journalist writing about gardening, I knew I could count on Jeff to be generous with his time and knowledge. Members of garden clubs have told me they also find him to be reliable, friendly and entertaining.

"I've never seen anyone create a garden so fast," an organizer at the San Diego Spring Home/Garden Show told me. "One minute it was an empty display space with a bunch of plants in pots nearby; the next minute it was a landscape." In addition to creating gardens large and small, Jeff is a blur when composing containers. One of my most popular YouTube videos shows Jeff creating two succulent dish gardens in fewer than five minutes apiece.

Inspired by the fact that many of the plants in his nursery look remarkably like reef inhabitants, Jeff created an under-sea-themed succulent garden for the San Diego County Fair. No one had seen anything like it, and it made him a horticultural celebrity. He later installed a permanent undersea succulent garden at the San Diego Botanic Garden.

The Philadelphia Flower Show—the largest of its kind in the U.S.—asked him to create a similar exhibit. So Jeff filled a truck with tentacled euphorbias, crested cacti and starfish-like aloes and drove to Pennsylvania. It was March, and it must have been a challenge to protect the plants from freezing. Philadelphia did San Diego one better: From the ceiling above his display, they suspended a boat. Some of the best-looking landscapes in my book, *Designing with Succulents*, show Jeff's own yard. He paints with plants: Blue *Senecio serpens* flows around clusters of red-leaved *Aloe cameronii* and orangey-green *Aloe dorotheae*. Adding contrast and texture are mounds of variegated elephant's food and spherical golden barrels.

As you're about to discover, Jeff is much more than a nurseryman and garden designer. He's a terrific storyteller with a delightful sense of humor. If you're looking for appealing ways to combine succulents in your garden, you'll find this book inspiring and chock full of helpful information. Most of the advice is specific to mild-climate regions of California. But regardless of where you live or how you came under the spell of succulents, you'll enjoy the beauty of Jeff's photos and the quality of his prose. His personality shines on every page.

— Debra Lee Baldwin

Preface

I hope you will approach this book the way some of us tackle a *National Geographic* magazine. Take a first run through and enjoy the photographs, perhaps stopping to read some captions and sidebars. Work your way back and read the text as time and interest permit. This book is intended to be more inspirational than educational, although I hope you learn something along the way. You can approach the world of succulents from a scientific perspective, artistic viewpoint, or as a casual gardener with an interest in growing them.

You will notice that this is a first-person perspective. As someone who has gone down the rabbit hole from casual hobbyist to serious collector, to nursery owner, I hope to shine a light on this world from an insider's point of view. These plants have a way of taking hold of your imagination. After thirty years of involvement, I still get as excited about a new plant as I did on the first day.

Echeveria agavoides 'Ebony'

It has been said that the best way to kill your enthusiasm for your hobby is to try and make a business out of it. I have no doubt that is sometimes the case; however, a flip side of that coin is the saying, "Love what you do and you'll never work a day in your life." I subscribe to the latter, although quite often it is hard work. I never tire of coming to work each day to deal with happy people buying things they enjoy. Plus, it is a life spent outdoors among some of the weirdest and most beautiful of nature's creations.

This book has a regional perspective. My nursery is in coastal Southern California, which has a benign to at least acceptable climate for these plants, which come from many corners of the world. Other parts of the world where succulents are easily grown are Africa, coastal South America, much of Australia, New Zealand, and the Mediterranean areas of Europe. The tropics can at times be difficult for some succulents—either too wet, or some are in need of a winter cool down. The biggest challenges are in places where winters are harsh—long and wet, freezing or frequent snow. There are some succulents that can handle the cold and I have been amazed at the quality of plants that are grown in upstate New York or Northern Europe. In these places, all you need is a heated greenhouse or garden window to keep a little slice of Africa or Mexico in your home.

A good portion of this book deals with growing succulents in the landscape. If you don't have that option, I apologize for tantalizing you with things you can't grow. If it's any consolation, even Southern Californians can't pull off the amazing tropical gardens of Florida or Hawaii. This book is intended for the enthusiast. There will be little about native habitats or conservation status. There are excellent books

Agave 'Blue Glow'

and journals that deal with those topics. If you're a beginner, this will be a window into a new world. If you've already slipped through that window, I hope to show you some things you haven't seen before, or perhaps renew your enthusiasm for your plant passions.

No book can come close to covering all types of succulents. There is a ten-volume tome on the euphorbia genus alone. The best I can do is offer some photos and information on the more frequently encountered groups of succulents, with a few rarities thrown in for inspiration. The interludes in this book will introduce you to some of the major genera. The chapters will address the different ways people engage with succulents, including certain groups of plants that share similarities despite being from different families, or even different continents.

The care and feeding of succulents will be dealt with in a series of questions based on conversations I've had with garden clubs and visitors to the nursery. I'll try to address all the

major concerns; believe me, after twenty years in the business, I've heard them all. Some answers are subjective, and you can get ten different answers from ten different growers. I'll try to offer what I consider the best-consensus opinion.

There is something depressing about looking at pictures of diseased or bug-damaged plants, and I consider it a waste of space to show you those at the expense of something of beauty. There will also be no photos of repotting procedures with the series of an empty pot, a tidy pile of soil, a gloved pair of hands and a plant ready to be potted. Again, this is a waste of valuable photo space. There are YouTube videos for all of that.

So, let us begin our journey into sublime succulence with a little foundation on just what constitutes a succulent and then dive into some of the varieties and a few of the ways people have fallen *Under the Spell of Succulents. . .*

Introduction

If you've ever had a hobby that revolved around a group of related items, whether things from the natural world such as rocks or minerals or butterflies, or collectible pottery or antique furniture, you probably had that first "light-bulb" moment where something caught your fancy. My journey into the world of succulent plants started with several bellwether specimens that caught my eye.

The first was a large variegated *Agave americana* that I biked past on my way to class at college in San Luis Obispo, California. It looked like some kind of a stationary yellow-striped alien octopus, sitting by itself in the middle of a lawn. I watched it throw up a 20-foot flower spike over the course of little over a month. What the heck was that thing?

I didn't quite have the motivation to research it, but it made an impression. Later, while eating lunch beneath a big Dragon Tree in Del Mar, California, I recall looking up and wondering what it was. Some kind of branched palm? A fat yucca? That pretty much exhausted my knowledge of succulents at the time, but I soon started to notice things while driving around town, particularly the dragon trees in the older parts of San Diego. These ancient monsters had been there all along, yet I had been oblivious to their stately existence. Then I saw an exhibit of bonsai succulents at the local fair, and it's been a downhill slide ever since: a life of appreciation, acquisition, disposition and fascination with succulents.

The display of bonsai succulents at the San Diego County Fair led me on a search of nurseries to find some elusive creatures, which led to more elusive creatures procured or desired, which led to a crowded patio of plants, which led to an experimental sale at a street fair, which led to more street fair sales, which led to opening a succulent nursery. I figured if I was this obsessive about succulents, there must be others like me out there that could keep me in business. Turns out there are.

Later in the book we'll talk about the collectors and the collectors' mentality. Suffice it to say that for those of us who are attracted to ornamental plants, succulents provide an almost mind-numbing diversity in form, color, and flower. For those with a brown thumb, succulents are often the first, and sometimes only, plant that you've managed to keep alive. That success can lead to another attempt, and a visit to the succulent section of your local nursery might launch a hobby. Once that window opens, it just leads to more windows and more journeys into these wonders of the natural world.

Succulents are primarily an ornamental pursuit. Some are edible, such as the opuntia cacti. Others, primarily *Aloe vera*, have medicinal benefits. Still others are sought for their psychotropic properties. But for those of us who covet our remaining brain cells, no hallucinogen is needed to enhance the incredible design and attraction of these architectural marvels of the plant kingdom.

The Dragon Tree, *Dracaena draco*

A group planting of *Aloe dorotheae* with blue *Senecio serpens* and *Portulacaria afra* 'Variegata'. Upper left corner is a large golden barrel cactus *(Echinocactus grusonii)*. This is a planting based on color, texture, and architecture of the plants, not on the basis of geography or genus. The golden barrel is native to Mexico, the rest of the plants to various parts of Africa. They would never occur together in nature, but get along great in any Mediterranean climate. Typical of most succulent plantings, once established, this tableau changes little over the years, as the plants are slow growers. Colors get more vibrant and the aloes bloom in the cooler months, but only very occasional weeding or thinning out of the senecio or portulacaria is required.

"Succulence" is just a description, not a genus or family. What makes a plant a succulent? Here is one way to think of it: If you stepped on it would it leave a wet spot? If so, it is probably a succulent. But I don't recommend that technique. The word "succulent" derives from the Greek word *succos*, or juicy.

Most succulents have evolved in dry climates where water retention is important. They store precious liquid in their leaves, stems, or roots, or some combination thereof. This gives most of them either a plump, or sculptural, or occasional grotesque appearance. After a while, you just know one when you see one.

Many plant families have certain members that have simply strayed into dry climates, or more likely, their habitats slowly dried up around them, and they evolved into succulents. For example, aloes are members of the lily family that have learned how to hold their water.

Some plants are "nebulously succulent." Plumerias are considered tropicals by some, succulents by others (hint: they leave a wet spot). Certain desert trees have developed fat, juicy bodies but are still somewhat woody, such as the commiphoras from Africa or the burseras from Mexico. By contrast, just about every cactus is extremely succulent: big balls of juice protected by vicious spines.

An assortment of tillandsias

Many succulent collectors also covet tillandsias (air plants), or other bromeliads, cycads, even certain xerophytic orchids. What these plants have in common is an other-worldly quality, as well as being sun-loving, low-water, and low-maintenance. Most enthusiasts tend to follow certain groups, or groups within groups that strike their fancy.

Here is an opening question I like to throw to the audience when I address a garden club: As the owner of a cactus and succulent nursery, what do you think my most frequently asked question is?

Do you get stuck by a lot of spines?

Yes, daily. I hardly even notice anymore. But that's not the question I'm fishing for.

Can you make a living doing that?

(Wise guy)...Well, I'm alive and talking about it, so I guess so. Let's try again. As the owner of a *CACTUS AND SUCCULENT* nursery, what do you think I'm always asked?

The difference between a cactus and a succulent?

That's the one. And the answer is?

A cactus has spines and a succulent doesn't?

Sort of true as a generality. If it has some serious spines, it's probably a cactus. But there are spineless cacti and some spiny non-cactus succulents. Any other ideas?

All cacti are succulents, but not all succulents are cacti?

Yes! Think of succulents as a huge and diverse group of plants. The cactus family is one very large group of plants contained within the larger group. Cacti are new world in origin; succulents grow worldwide.

Echeveria cante

I was told you are only supposed to water your cactus when it rains in the desert, so I look in the paper to see when it's raining in Arizona to water my plants. Is this right?

Well, considering most cacti and succulents in cultivation are from Africa, Mexico, or the Canary Islands, I would only apply that method to any plants you knew were from Arizona. And even then, your climate is probably very different from the desert Southwest. The best general answer as to when to water a succulent is "when it is dry." And to make it easier, I usually try for once a week, adjusted a little more frequently in hot weather and little to none at all in wet winter conditions. Go ahead and water thoroughly, until water comes out of the drainage hole (providing it is planted in cactus soil—more on that later). In nature, many succulents spend most of their lives shriveled and desiccated, hoarding their water in tissue protected from the sun by its particular method of sheathing. When it gets the rare opportunity to drink rain, it will puff out in all its glory, proceed to flower and fruit, to once again retreat for months into its guise of leathery ugliness. In cultivation, we tend to baby our plants, and for good reason: it is an ornamental appeal we are after, and who wants to look at a pathetic, shriveled plant? Fortunately, most succulents are happy to spend most of their lives pampered and they will luxuriate in our care. There are exceptions: lithops (living stones) are one of the plants most likely to be killed with kindness. Water them too much out of season and they implode. Water them too much during growing season and they explode. This sort of defies the low-maintenance appeal of succulents, but we still try, because c'mon, they're living rocks!

Heart-shaped opuntia

Cleistocactus

Echinocactus grusonii (foreground) and *Agave colorata* (background)

Agave victoriae-reginae compacta

Senecio rowleyanus, or "string of pearls" ("string of peas"?) is one succulent that may burn in too much sun; try morning sun only at first, or filtered light. The flowers smell like cloves. There is also a pearlescent pale variegated form occasionally available.

Is tap water OK for my plants?

Generally it is, but it's not as good as rainwater. No matter how much I hose-water my in-ground succulents, there's nothing like a good first rain to bring some things to life. Some succulents (fouquieria) just seem to recognize rainwater. Unfortunately, so do the weeds.

Here's a technique to improve the quality of your tap water that has been used by bonsai enthusiasts as well as succulent growers. Increase the acidity of tap water by adding a teaspoon of distilled white vinegar per 5 gallons of tap water (most succulents like an acidic vs. alkaline water). Some of us will also add ammonium sulfate in trace amounts, and an occasional drop of plant food. We're getting scientific here, which goes against the succulent grower's "sun/water once-a-week/cactus mix soil" credo, but this is a technique that you can research further. It is practiced in some form by many expert growers, but you need to first check the acidity of your tap water and adjust for a pH 5.0 to 5.5. I hope I haven't sullied your succulent experience with too much science.

What types of soil do I plant succulents in? Sand?

You would think sand would be at least a major soil component, as that seems to be what desert soil is made of. Often in habitat the plants are either growing in sand or decomposed granite with perhaps some kind of organic loam. However, that doesn't seem to work very well in pots. The sand gets wet and heavy and stays that way too long. Most growers use some form of a "cactus mix" or "cactus and succulent mix." Sand may be a five or ten percent component, but almost always, the magic ingredient will be the little white pebbles, which are either gravel-sized natural pumice or its man-made equivalent, perlite. This "white stuff" tends to lighten and aerate the mix, while also storing some water in the micro-pockets for the roots to get to.

Different growers will debate which is better (most go with pumice), and many will use a combination. A good mix will usually be at least 50% pumice and/or perlite, with the remaining half mostly some type of brown organic planter mix or compost. There may be a small percentage of sand, and occasional trace elements of nitrogen, etc. depending on the sophistication or bias of the user. Most nurseries will offer an acceptable cactus mix. Check to see that it has some "white stuff" in it.

When planting in-ground, perlite as a soil component isn't recommended, as it tends to float to the surface and make little snowdrifts around your garden.

Can I grow my succulents indoors?

Only in a sunny window. Some will survive with bright indirect light, but think of them as desert plants. A minimum of a few hours of direct sunlight is best, with mature specimens generally preferring to bask in the sun for at least half a day. If the summer daytime temperature gets to triple digits, there are many succulents from more temperate climates that will burn. Slowly acclimate them to as much sun as they can handle. A few succulents that tolerate lower, yet still bright or indirect light include most haworthias and the Christmas cactus, schlumbergera. Others, such as *Euphorbia trigona* or *Beaucarnea recurvata*, can adapt to lower light and live for years, albeit in

An example of the three R's of succulent garden design: Rocks, Restraint from including too many varieties, and Repetition. The winding blue swaths of *Senecio serpens* and the green flowing aeoniums give this garden a pulled together and natural feel, and allows the specimen golden barrels and aloes to stand out. Design by Michael Buckner.

notocactus

more of a stretched-for-light manner. They would prefer the sun, but can adapt to less.

Some succulents may initially burn in full sunlight, but only because they may have come out of spending an extended time in a shaded part of a store, or a greenhouse with filtered light used to bring it up from its early stage of life. Gradually acclimating a plant to more and more sunlight is recommended. Direct sun tends to bring out the best colors in most succulents, although you will get some arguments from many serious growers, who recommend either shade cloth or greenhouse glass to prevent extreme sun from burning some plants. Indeed, some of the most amazing "show plants" at cactus club shows spend their lives under these conditions. I prefer to grow plants "hard," which means exposed to the elements, for better or worse. Do some research to see where your plant is from, which might give you an idea how much of your local weather conditions you can expose it to. If you really want to keep a succulent in your artificially lighted office, at least try putting it in a weekly rotation back to semi-sun. If that is too much work, you can look at them as replaceable plants; they can take forever to die. But that is too cruel. Manufacturers are getting really good at making silk and plastic succulents. You can even identify some by genus and species.

Most succulents, and especially those highlighted in this book, are considered more Mediterranean than desert plants. This means they generally tolerate lows into the 30s and highs into low triple digits. Some (aloes, aeoniums, most crassulas) are OK with cool and wet winters. Others can handle a dry cold, but not a wet cold. Many coastal plants, such as those from the Canary Islands, will flourish within say 10 to 20 miles from the coast, but summer inland heat or winter frost might do them in. They may do fine for a number of years, but that one 5- or 10- year frost, or heat wave, might finally kill them. Many cacti grow and flower much better 10 to 20 miles away from the coast. Some plants adapt easily to their new home-away-from-home, others never get over their homesickness. Try to get the best information you can from the nursery you purchased your plants from, or do your own research. Sometimes experience is the best teacher. Serious succulentoholics will usually have shade cloth structures, or greenhouses, or both. I've killed enough adeniums and *Aloe polyphylla* that I should know when to quit.

Should I ever fertilize my plants?

Yes, just don't overdo it. Any household plant food with numbers in the 10-20 range is fine, but at half strength or less, and only on occasion. Overfeeding succulents goes sort of contrary to their lifestyle, and can have a steroid effect that ultimately weakens them, although at first they may look great. Some "cactus" fertilizers are just the regular stuff pre-diluted, so just dilute it yourself. Some growers prefer granular time-release or fish emulsions; again, just experiment, but err on the side of going light. This is also "growing hard." Some growers like to push their plants with food to get them up to size fast, but just like a juiced-up body-builder, they will suffer and sag eventually. They live a harsh life in their native habitats; try for something slightly more pampered.

Some succulents make a case for simply being beautiful. This cluster of *Aeonium* 'Sunburst' (facing page) appears to be a bed of full-time flowers. Even those who say they can't stand cactus will usually admit that this type of succulent is OK. There is a succulent for everybody!

A *Carunculated echeveria* displaying a sort of desired plant leprosy.

Succulents have developed myriad and sometimes mutated forms to survive in harsh environments. It can be argued that no single collection of plants have such varied forms and appearances, yet somehow appeal to us as a loosely connected group of beauties and beasts. The plants here will attract those who appreciate the macabre or science fiction. Some just don't seem to be of this earth, beautiful in their ugliness.

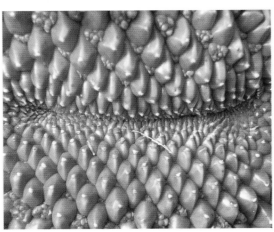

Clockwise from top left:

Sempervivum arachnoideum crested form; *Faucaria tigrina*, tiger's jaws; *Lithops aucampiae*; *Myrtillocactus geometrizans* crested form; *Euphorbia flanaganii* stem crest; *Echeveria*, frilly hybrid. The faucaria looks like an alien about to eat you; the myrtillocactus like a tightly stacked series of river rocks. The lithops like a . . . umm. . . maybe a plumber with a case of the hives?

Kingdom: Plantae

Phylum: Viridae

Class: Equisetopsida

Order: Asparagales

Family: Xanthorrhoeaceae

Genus: Aloe

Species: arborescens

Aloe Ferox v. candelabrum, with early flowers

Aloes

After 20 years in the nursery business, you can tell what my favorite type of plants are by inspecting my yard. Aloes everywhere. They are primarily native to Africa, but most feel right at home in Southern California. Shortly after opening the nursery, I came to appreciate some of the sculptural branched and tree aloes, but hadn't yet developed an eye for the whole genus. Then an enthusiastic customer bought some strange ones that came in with some other plants, and made an offhand comment that he "needed" *Aloe broomii.* I thought it was funny, his use of "need" rather than "want." Then I delivered them to his house. I hadn't yet seen a collection primarily devoted to a group like he had been working on. It was winter, and there were aloe flower spikes in various stages, from common red to bicolor and branched, yellow to orange and in-between. And it wasn't just the flowers. The reptilian skin of some aloes were tinted pale burgundy;

Aloe krapholiana

others in vibrant greens or with variegated yellow stripes following a lazy spiral from the center. Some were large multi-headed clumps, others solitary rosettes on vertical trunks, and still others looked like strange trees from a Dr. Seuss book. Yet they were all aloes, their kinship apparent. I got it. He needed *Aloe broomii.*

The aloe compulsion took hold of me as well. Many are available in California, but books with habitat photos of rare and unavailable aloes give the collector something to shoot for. I could tell a serious aloe collector from his list of needs. *Aloe pillansii* or *Aloe pearsonii* are among the "Holy Grail" plants to complete a collection.

I have since worked through the compulsive aloe collector phase, although my yard offers mute testimony to this indulgence. I'm out of room and rarely bring a new aloe home anymore. However, I recently came across a new hybrid, a cross of *A. capitata* and *A. munchii.* I bought two from a collector; one to sell at the nursery, but the other I kept. I had to. I needed it.

Aloe secundiflora

Aloe arborescens, spineless form (stressed red)

Aloe cameronii

Aloe africana flower

Aloe speciosa flower

Aloe hybrid

Aloe hybrid

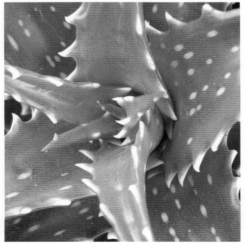

Aloe dorothea

Aloe polyphylla

31

Aloe rupestris

Aloe hybrid 'Tangerine'

Aloe claviflora

Aloe capitata

Aloe tomentosa

Aloe flowers are universally cool to behold. They start off looking like little lizard tails poking out of the plant, keep growing, and end up as either single or sometimes branched candles of color and iridescent glory. Aloes bloom primarily from late fall through early spring. There are a few repeat bloomers, or summer bloomers, such as the fuzzy flowers of *Aloe tomentosa.*

FACING PAGE *Aloe rubroviolacaea*

Aloe arborescens (California's Christmas succulent)

In the introduction I said I didn't want to show you pictures of bug damage, but I suppose we can talk about it. Aloes can get galls caused by microscopic aloe mites. It is non-fatal but ugly and can spread. Try to cut out and dispose of the infested leaves or flowers. If it comes back, there are some systemics that can kill them, and I've been told bleach in a spray bottle might work. Aphids can be a lesser pest, carried about via ants. Assuming the aloe is sturdy enough, first blast out the invaders with hose water. Then spray with any off-the-shelf insecticide that targets aphids. You may need to repeat the process a few times over a few weeks.

Aloe vera

Aloe tongaensis, formerly known as the dwarf or Medusa form of *Aloe barberae (bainesii)*. It is an excellent landscape specimen, set apart from the larger *A. barberae* by more profuse branching and orange flowers. Easily propagated by cuttings, this specimen has been trimmed of its lower growth both to accentuate its tree-like nature, and as a stock plant for a future forest of its clones.

Do all aloes have healing properties?

Most contain the chemical "aloin," a gel that is useful for treating burns, as well as being used in certain internal medicines and cosmetics. Almost all aloe gel, however, is generated from *Aloe vera*, the medicinal aloe. It apparently has the best concentration and is the easiest to mass-produce. When you refer to *Aloe vera*, you are referring to only one of over 500 species of aloe. It is a somewhat generic-looking but easy-to-grow plant, usually with a nice yellow flower. There is a great deal of evidence that you can soothe and heal burns and cuts directly from the gel applied from a severed leaf. It is one of the few plants I sell for reasons other than ornamental. *Aloe arborescens,* by far the most common and prolific ornamental aloe around the world, likely also contains beneficial aloin. As you can see from the photo on page 34, *A. arborescens* is an ideal candidate for a succulent fence, or in this case, absorbing and becoming the fence. As with most aloes, it looks at first blush to be a spiny "stay away" plant, and serves that purpose, but in fact the teeth are soft and won't draw blood. And even if it did, well, it's an aloe, so it would automatically be providing its own first aid salve, right?

ALOES OFFER BOTH STRUCTURAL AND FLORAL APPEAL

TOP LEFT is the flower of *Aloe hildebrandtii*.

TOP MIDDLE is the hybrid *Aloe* 'David Verity', a blue aloe with a bicolor flower.

TOP RIGHT is *Aloe striata*, the coral aloe.

ABOVE is the fan aloe, *Aloe plicatilis*. It is a dramatic piece of architecture for any garden.

AT RIGHT is *Aloe vanbalenii*, an "octopussy" red-blushing aloe.

Photos this page by Bob Wigand

This is likely a hybrid form of *Aloe ferox*. Regular *A. ferox* or *A. ferox v. candelabrum* usually has a redder flower. Some aloes flower with single stalks, while a handful will flower with impressive elk horn or candelabrum flowers such as this. After the flower is spent, the dried flower stalk makes an interesting ornament, much as the dead flower stalk of an agave. Check any dried pods for seeds before they open and blow away. A single-head aloe can only be propagated by seed. A frequent result of growing from seed are hybrids, especially if seed is collected from openly-pollinated plants outdoors. Some collectors have no use for hybrids, but unless you are planting a true-to-habitat botanic garden, have an open mind about hybrids. Some are more vigorous and more prolific flower producers than their parents. Excellent hybrid aloes include *A.* 'Cynthia Giddy', *A.* 'David Verity', and *A.* 'Hercules'.

I have two aloes with the same name but they look different. Is one mislabeled?

Possibly, but there can be variability within the same species, or an accidental cross-pollination may have occurred at some time in the past. For example, there seem to be several versions of the red *Aloe cameronii* in cultivation. They may be representative of actual habitat variation in collected starter plants, or subsequent cross-pollination. Sometimes they are referred to as different 'clones' of the same plant. This happens with many if not most groups of plants. Other times mutations are isolated and become their own distinct cultivar. Variability is what can make life interesting. *Vive la différence!*

TOP LEFT *Aloe ramosissima*. This aloe is similar to *Aloe dichotoma* in its needs (see below), and is closely related. It differs in its copious branching, giving it more of a bushy look.

MIDDLE LEFT *Aloe ramosissima* in late fall bloom.

ABOVE *Aloe dichotoma* is a tree aloe from the dry regions of Namibia, where it gets some winter rains and endures long, hot dry summers, much as inland southern California experiences. This is one of the "don't water" aloes; too much can rot it. Plant it high and dry and give it very little attention. It is a very slow grower. The hybrid of *A. dichotoma* and *A. barberae* is known as *Aloe* 'Hercules', and will accept water and grow much faster. *Aloe* 'Hercules' is shown in the later chapter on succulent giants. *A general rule on most cultivated aloes:* The more glaucus (blue-gray) plants prefer hot and dry conditions; the greener ones seem more thirsty and tropical/cold-sensitive.

BOTTOM (AT LEFT) The much more rare (in habitat as well as cultivation) *Aloe pillansii*, likely a regional sub-species of *A. dichotoma*, also known as the giant quiver tree, is shown at far left, while *Aloe dichotoma* is shown in bloom at the San Diego Zoo Safari Park (behind *Aloe pillansii*).

DWARF ALOE HYBRIDS

There has been a lot of interest over the past few years in miniature aloe hybrids. Hybridizers Kelly Griffin, Karen Zimmerman and others have been creating new hybrids, following a line of crosses started by John Bleck and Dick Wright several decades ago. Original source materials included *Aloe descoingsii, A. juvenna, A. humilis,* and other small species. The new hybrids are far removed from anything that exists in nature, and are a testament to how far we can push things. New colors, patterns, teeth dentition, and to a lesser extent flowers are the goal. The original hybrids of *A.* 'Lizard lips', *A.* 'Cha Cha', and *A.* 'Doran Black' have given way to aloes named 'Christmas Carol', 'Chainsaw', 'Gargoyle', and 'Sunrise'.

Aloe chabaudii

Aloe rupestris

Aloe glauca v. muricata

Aloe ferox flower

Aloe collectors can usually come up with a top 10 list of landscape aloes. While these may not make a collector's top 10 list, which usually has some rarities, I'd nominate the following in each category:

Centerpiece/sculptural: *Aloe barberae, A. tongaensis, A. 'Hercules'.*
And if you are careful not to overwater: Aloe dichotoma, A. ramosissima, A. plicatilis

Colorful clumpers: *Aloe cameronii, A. dorothea, A. elgonica*

Statement plants with "wow" blooms: *Aloe speciosa, A. marlothii, A. rupestris*

Smaller repetition "garden" aloes: *A. rooikappie, A. 'Cynthia Giddy', A. brevifolia*

Hedge or fence aloe: *Aloe arborescens* or one of its many hybrids.

That is way more than 10 plants! Hey, I'm an aloe guy.

Haworthias & Gasterias

Haworthias and gasterias are close allies of the aloes, and can in fact sometimes be crossed with aloes and each other. In general they are smaller and better suited to container gardening, although most do fine in-ground. Flowers are usually not as showy, although some gasteria flowers look like little inflated flamingo heads or pumpkins on a stick. As with many succulents, some haworthias and gasterias can provide stunning examples of variegation and can be highly valued by collectors. The small size and lower-light tolerance of haworthias in particular makes them well suited for windowsill gardening in less hospitable climates.

Haworthia fasciata

Haworthia retusa

Gasteria hybrid

Haworthia cuspidata (?)

Haworthia emelyae

Haworthia emelyae

Haworthia cooperi

Haworthia truncata 'Lime Green'

Gasteria bicolor hybrid

Haworthia tesselata

Haworthia limifolia variegated

Gasteria glomerata flower

CHAPTER 01: Container Gardens

Succulents are among the most appropriate plants for container culture. Due to climate, or limited (or no) yard space, many people have an exclusively potted plant collection; and some succulents do best, or just show best, in pots.

My potted succulent looks pathetic. What is wrong with it? Am I watering it too much or not enough?

Not really enough information here, but there are a few things to consider first. Are you watering when the soil seems dry? Again, a thorough soak every week or two, with an occasional dollop of dilute plant food is a rule of thumb. Is it planted in a good cactus/succulent mix (see page 22)? Is it getting enough light? How many years has it been in the same pot? If you can't remember the last time it was repotted, maybe it's time. Intuitively, if a plant has been in the same soil for a long time, it may have depleted any last nutrients and will benefit from some new organics to work with. One commonly encountered malady is what I call "yellow-cactus syndrome." Assuming it's not a variegated plant that is supposed to be yellow, the plant is stressing into a jaundiced state. There is no single cure-all diagnosis or remedy, but the steps to take would be to first repot with new cactus soil, then try fertilizing (not too much), watering with the vinegar-adjusted pH as discussed on page 22, then adding a supplement such as Ironite. Then try talking to it.

Having said that, I've had plants in my collection in the same pot for over 15 years, and I've resisted repotting because they still look good. A lady once brought a *Beaucarnea recurvata* (ponytail palm) to my nursery to be replanted. She inherited it from her grandmother, and it looked like a bowling ball in a cereal bowl. I broke the pot off (the terra cotta was in the process of being absorbed into the plant), and there was literally no soil left, just a saucer-shaped root ball. And still the plant looked fine, with long green leaves, still soaking up water. I never saw it again after repotting, but with the larger pot, I'm sure it was thrilled to finally break out of its stunted life and grow (check out the beast on page 216 to see what it can turn into).

Some nicely grown-in succulent containers softening up a hardscape. Just when you reach maximum full-plant-glory, it is about time to consider thinning out. It can become a challenge just getting water into the pots when they are this full. Your plants will tell you when it's time by shriveling and drooping. You can just heavily prune them back, or empty the pots, add fresh soil, and put about half of the old material back in, either as un-rooted cuts or the original rootstock.Either way, you've doubled your inventory. As a nurseryman who sells plants, it hurts to advise you on how easy this is.

This large "cactus" is not a cactus, but *Euphorbia acrurensis*, thriving in a pot that compliments its background. Euphorbias can live for many years in a pot, but long to be liberated into the garden. Photos on pages 44 and 45 by Bob Wigand.

TOP ROW (left to right) *Aloe cameronii, Aeonium* 'Cyclops' with *Euphorbia* 'Sticks on Fire' along with some cascading pot-mates, *Parodioa magnifica*.

ABOVE A "cornucopia" bowl of abundant succulent life. This one features colorful aeoniums, echeverias and crassulas, providing year-round color. Design by Tita Heimpel of Courtyard Pottery.

AT RIGHT Succulents can liven up any flat space where you can set a pot.

A densely planted dish garden of soft pastel succulents, including *Sedum nussbaumerianum*, *Kalanchoe tomentosa*, *Crassula platyphylla*, graptopetalums and sempervivums. They crowd together nicely, their colors playing off each other. As long as you combine plants that appeal to your sense of color compatibility, and that grow at similar rates in the same conditions, why not build an explosion of abundant succulent life?

There are several reasons why people get into succulents. The best way is through fascination and a desire to collect or learn more, and the enjoyment of variety and shapes. As a plant guy, that is my preferred path. However, some folks have a patio that they want to liven up, and they've decided to do it with succulents, likely due to the low-maintenance aspect. Once they've got it planted, it's done and they move on to the next thing. Hopefully they find time to stop and soak in the beauty. And lastly, there are those who have a few succulents left in their pots due to attrition: everything else has died. Contrary to popular wisdom, succulents don't thrive on neglect, but many can handle it beyond all expectations. "Benign neglect" is a better aim; just take a time-out on occasion to remember they are living things and can't be ignored forever. The flip side of that is to over-mother them. Too much water, especially if they are in lower light and can't dry out between waterings, may eventually take its toll. Remember, water when dry. That usually works out to once a week-ish if they're in strong light.

ABOVE A baking-hot hardscape can be softened up nicely with potted succulents, or in this case, succulent troughs. Here, the *Euphorbia tirucalli* 'Sticks on Fire' glow against a blue wall; variegated *Portulacaria afra* cascades to break up the edge; and the *Euphorbia milii* 'Crown of Thorns' provides year-round flowers. If you have a white stucco wall, consider painting it a vibrant color. It will make you happy.

TOP RIGHT *Fenestraria aurantiaca* (baby toes) is one of the more popular mesembs in cultivation. They can be winter-sensitive in the landscape, but in containers you can control their environment. They like to watch the winter wetness from a dry windowsill.

MIDDLE RIGHT They dream of Africa (except maybe the cactus in the middle. He probably dreams of Mexico). Nevertheless, these little plant pets are happy enough in a sunny window year-round, especially during cold and wet or snowy winter months. Direct sunlight a few hours a day is best, as these have it, but some will tolerate bright indirect light. They may lose color or get leggy in lower light.

BOTTOM RIGHT Lithops, conophytums, and pleiospilos (living stones) in a pumice planter. Rocks in a rock. Organic rock planters allow you to plant succulents in a manner that often replicates how they grow in nature. Photos of their native habitat often show them growing in the cracks between rocks, or in the case of lithops, among a group of similar-sized rocks and pebbles. They announce their presence only while in bloom, then blend back into their rocky habitat.

ABOVE Another "non-spiny" pastel color bowl. Low bowls are excellent for multiple-plant dish gardens. Try to mix a plant with some verticality in back (here it is silver dollar jade), with colorful rosettes such as *Echeveria* 'Afterglow', and a cascading plant like the donkey tail *Sedum morganianum*, also known as burro's tail, or sometimes as *Sedum* 'burrito'.

TOP LEFT Cascading succulents such as these in trough planters can really liven up a balcony. Assuming the plants were 6-inch starts, it would take a year or two to get such colorful abundance. Highlighted here are *Sedum morganianum*, *Crassula* 'Campfire', cotyledons and crassulas.

MIDDLE LEFT A trough planter of *Aeonium* 'Kiwi', which in this case shows both the variegated yellow form and reversions to green; both look nice. If a pot or trough sits flush on the ground as seen here, be sure the drainage holes don't clog. Even if planted in cactus soil, succulents can rot in wet winter months if the soil turns to soup. You can place small pieces of broken tile at the corners to elevate it just a fraction of an inch off the ground to allow drainage.

BOTTOM LEFT *Rebutia muscula*, a superior white fuzzy mounding cactus in maturity, presents wonderfully in this hand-built organic-looking stoneware pot at a cactus show.

TOP ROW (far left) If you have a $100 hand-built pot, it deserves a $100-plus plant. Here is a valuable *Dioscorea elephantipes* dressed up in a showcase piece of art. A bonsai maxim is that a plant must earn its pot. An older caudiciform like this deserves proper presentation. **(Middle photo)** A pedestal planter nicely overgrown with a cultivar of *Crassula perforata*, with *Kalanchoe luciae* at top. This arrangement could feel at home in an "undersea" succulent garden (see page 140). You can tell which way the current is flowing. **(Far right)** *Sedum nussbaumerianum* in a hanging glazed pot.

ABOVE A giant clam full of colorful succulents. Treat these crowded cornucopia arrangements as long-term living bouquets that will eventually either have to be thinned out or redone when they begin to look tired or just too ridiculously crowded. This clam shell is worth way more than the plants, so don't drill a drainage hole. It likely is shallow enough to allow evaporation. Manufacturers also make some pretty convincing artificial shells.

Succulents are generally shallow-rooted, and the plants in the above photo really only require a pot of maybe one-third the depth of these vessels. The bottom half or two-thirds can be gravel or even Styrofoam peanuts, or fill with cactus soil all the way up. Another option is to find a plastic pot that just fits inside the rim, and elevate it on top of an inverted pot inside so the top of the containers appears to be directly planted. The planters above are along a north wall, and for at least four months in the winter, they receive little to no direct sun. This is not their preferred light, but they can handle it as long as they eventually recharge their solar batteries spring through fall.

Be sure your pot has adequate drainage. Drainage holes are mandatory, unless you have an unusually shallow pot. Be sure the drainage holes are large enough not to clog with roots. You can use a cut piece of screen or drywall tape to prevent losing soil through a large hole. Or you can use several layers of newspaper to cover the hole. By the time it decays, the soil/roots should keep the mix intact in the pot.

A tightly-stuffed "glory bowl" ready-to-go from your local hardware store. The presentation looks great, small-in-front, large-in-back, nice color and architecture. But be careful planting a bowl too full. Plants need to grow. This arrangement will grow itself right out of this pot in a hurry, particularly the Senecio vitalis in back. In six months you could probably extract the entire conjoined mass of plants into a bigger pot.

TOP ROW (left) Entryway beautification with succulents. It is easy to get carried away, but as long as you can still get to your door... Plant lovers can have a hard time containing their enthusiasm for containing their plants. It can be frustrating if you have a small apartment with limited space, but a collection of succulents can really beautify the space you have. Read the upcoming chapter on collectors and see if you can relate.

TOP ROW (right) A dish garden should feature plants that grow at similar rates. Try for at least one with some verticality, another that will cascade or soften the pot's edge, and some rocks to give it the feel of a miniature landscape.

MIDDLE ROW (left) This turtle is shouldering a heavy load of euphorbias and haworthias. Like the wall pot above middle, this container is packed tight with overgrown plants, yet still looks great. Do your best to do a thorough watering if you can find a way to get it in. You can submerge a small pot into a bucket of water if all else fails.

MIDDLE ROW (middle) Wall-mount planters need cascading plants like the graptosedum cross seen here. It can get increasingly difficult to water overgrown pots like this because the plants start to crowd each other out. A quick spray with a hose might not be effective. It is a good idea to keep a watering can handy so you can do a few slow pours on occasion and wait for the water to seep in. Water until it leaks out the drainage hole, at least on every other watering.

MIDDLE ROW (right) A wide, shallow bowl of succulents can enhance a pillar or column. Here an *Agave* 'Red Margin' is featured, with a cascading *Portulacaria afra minima*.

INSET RIGHT An exceptional specimen of *Alluaudia humbertii*. There are only a handful of alluaudias in cultivation, *Alluaudia procera* being the most common. Apparently *A. humbertii* is a rather unremarkable, bushy plant in habitat, but we can make it look like a winner with the right combination of pruning and container. Like most alluaudias, it will defoliate in cold weather.

The Madagascar palm, *Pachypodium lamerei*, conveys the feel of a mutant palm from some faraway place. In fact, it was featured as a houseplant on the Starship Enterprise; hence the occasional name "Trekkie tree." It is a plumeria relative with white flower clusters when it reaches maturity. It's usually ten or more years old and several feet high before its first flowering event.

Euphorbia bupleurifolia

Face pots by Shannon Applegate

Presentation or "staging" can really compliment your prize plants. For some people, the plant is the thing, and they keep their collection in plastic or basic terra cotta. I have quite a few plants at home in all manner of unimpressive pots. However, if you have what you consider an important specimen, perhaps a show plant, consider staging it in a hand-thrown or hand-built pot, or perhaps an Asian bonsai pot if appropriate. Be sure the pot has adequate drainage holes. There is some controversy about the breathability of high-fired stoneware vs. terra cotta, but I've found either will work provided there is sufficient drainage. Hand-built stoneware (as seen on this page) is art in-and-of itself. I still haven't found the proper plants for some of my own Don Hunt pots, but they look fine as an empty collection of art until the right plant comes along and earns its pot. Lest we get too snobby, ceramic cowboy boots or coffee cups are fine if kitsch is your thing. Or perhaps you have a decommissioned pickup truck or rowboat that you can have fun with. But let's draw the line at cottage cheese containers. Whenever possible, support your local potter; most of them aren't getting rich but love their art!

55

Flowers of *Echeveria pulvinata* 'Red Velvet'. This plant is known for its stunning fuzzy/burgundy leaves (at right), but also puts on quite a flower show in the early summer. The flowers have the illusion of an almost incandescent glow.

You can plant strawberries in a strawberry pot. Or you can have fun with succulents. This recently-planted white glazed bowl is full of fuzzy echeverias, primarily *Echeveria pulvinata* 'Red Velvet'. After a year or two the plants will cover much of the pot. If they get leggy, just cut them back and wait for them to grow back out. It will be ugly for a while, but the good news is you will have a bunch of new cuttings to start elsewhere.

Another look is to have three or four different plants in the various pockets, and perhaps a large vertical plant, such as an aeonium or aloe, in the top. Be sure you have at least a few cascading plants. The best way to plant a strawberry pot is to fill soil to the first pockets, plant those, then fill higher, plant again, up to the top. If you are starting with all bare-root cuttings as seen at right, you can fill it all the way up and simply stick the cuttings in and wait.

The white-glazed strawberry pot (at right) was filled with *Echeveria pulvinata* 'Red Velvet' cuttings, and this same pot in bloom only 4 months later is in the top photo.

Some succulents have a miniature feel and can represent small versions of larger plants. You can create a "bonsai" dish garden of succulents, such as the church garden above. The ceramic Mexican church gives a sense of scale to the plants. The "tree" behind is *Crassula sarcocaulis*; the "cactus" is *Euphorbia ferox*. The echeverias in front suggest agaves, and the lime-green plant (*Deuterocohnia brevifolia*) suggests a clump of *Aloe arborescens*. Broken pieces of flagstone serve as the hardscape. This recently-planted garden will fill in nicely over a few years, but will eventually need thinning out. As the church decomposes with exposure to the elements, it will take on a truly rustic and authentic look.

Kingdom: Plantae

Phylum: Streptophyta

Class: Magnoliophyta

Order: Rosales

Family: Crassulaceae

Genus: Aeonium

Species: A. arboreum

various forms of Aeonium atropurpureum

Aeoniums

Aeoniums are indigenous to the Canary Islands, which means they feel at home in any benign coastal climate, luxuriating in fog and rainy conditions, and contracting and withering in the dry heat of summer when grown inland. They can be described as rosettes on stems, usually offsetting to form clumps or small upright shrubs. They enjoy wet winters and tend to look their best from fall through spring, when they will sometimes flower with large yellow or cream-colored spikes. An entire rosette will elongate into the flower, and by summer that portion of the plant will brown and wither, to be replaced by the growth below it. Aeoniums usually shrink over the summer, even with watering, but will almost always regain their full look by winter. They are excellent in the landscape or in pots. Favorites include the black rose, *Aeonium* 'Zwartkop'; variegated *A.* 'Kiwi' and *A.* 'Sunburst'; and the large dinner plate aeonium, which may be a form (or hybrid) of *A. canariense* or *A. urbicum*. There are several newer hybrid aeoniums popping into cultivation, usually in varying degrees of burgundy to black.

Aeonium tabuliforme

Collectors pine for the extremely flat *Aeonium tabuliforme*, which resembles a spiraling lily pad, but it can be difficult to grow. In habitat they look like green tortillas plastered vertically into cracks of rock cliffs.

Aeonium pseudotabuliforme

Aeonium 'Sunburst'

Aeonium 'Zwartkop'

Aeonium canariense hybrid

Aeonium canariense hybrid

Aeonium canariense

Aeonium 'Cyclops', a giant cultivar that can grow chest-high.

Aeonium canariense, stressed red

An unidentified aeonium hybrid. Possibly A. cultivar or a cross of A. ciliatum with A. davidbramwellii.

One of the newer hybrids, likely an *Aeonium simsii* hybrid called *A.* 'Cabernet'. It shows excellent fullness and color and will make a wonderful bushy garden plant; however, like many of its kin, it suffers during hot, dry summer months and tends to wither dramatically, only to puff out again after the first fall rains.

An observation on aeoniums: many aeoniums in cultivation, and in particular the black forms of *A. atropurpureum* or *A.* 'Zwartkop', may lose their appeal with age. After a number of years, they can get leggy, with just small compact rosettes at the end of long stems. While they still may fill out during their winter growing season, they tend to reach a point of no return to their former glory. This look has its own appeal, but most folks prefer the fat and bushy look. You can decapitate the heads and start them over with a few inches of stems, and discard the remaining in-ground stalks. Or support your local nursery and buy a new one, perhaps a variety you don't have yet.

One of the older, long-established aeoniums in cultivation, *Aeonium arboreum*. It truly is at home in Southern California, and will thrive in vacant lots or coastal cliffs and hillsides.

Do succulents get bugs?

Sadly, they do, just like all plants. Some seem almost impervious to pests, others are more of a magnet. Conditions play a factor: bugs seem to spread faster in enclosed or indoor situations, and are less of an issue with plants growing in the sun. And a healthy, growing plant is less prone to infestation than a weaker plant. But anyone who owns plants has to deal with them. Typical pests are mealy bugs (the little cottony gooey things), aphids, spider mite, and the ones I hate most: scale. Scale insects are like barnacles. They may have to be treated with a systemic insecticide, picked off one at a time, or in bad cases, you have to dispose of the plant. The other pests can be treated with anything off the shelf that includes them on its list of offenders killed. Some pesticides are citrus- or oil-based; others have some amount of poison. Most plants need several treatments over several weeks, with some wash-off in between. Diluted rubbing alcohol on a Q-tip will work if you have the patience, or some combination of alcohol and soapy water.

Aeonium 'Zwartkop'

What do you like for killing snails?

I prefer a 9-iron for loft to get them over the fence. Or toss them on the street. Or snail bait. There are mammalian nibblers as well. Gophers, rabbits, even deer may take a liking to something in your garden. Buried chicken-wire baskets surrounding the root ball will help if you have particularly aggressive subterranean vermin.

Aeonium holochrysum

Aeonium 'Black Night'

Several forms of green aeonuims provide low-water green year-round.

Aeonium 'Blushing Beauty'

Flower cones of the *Aeonium* 'Zwartkop'

Aeonium nobile

Aeonium 'Starburst'

Aeonium 'Starburst' is the inverted variegated form of *A.* 'Sunburst': a yellow interior stripe surrounded by green. *Aeonium nobile* is easy to grow but hard to propagate; hence it is rarely available. Most aeoniums, like many succulents, can simply be divided because they produce many heads. But in the case of a usually single-headed plant, such as *A. nobile*, the only way to make more is by leaf propagation (notable with kalanchoes), or from seed. *A. nobile* will eventually morph into a huge red flower cluster, and then die (it's mono-carpic, like agaves). The only hope for more is slow and challenging seed collection and propagation. Some brave growers know how to "core," or dig out, the growing meristem, in hope that the plant responds by forming multiple little rosettes...that is, if it doesn't die. It's nice that some things are still rare.

CHAPTER 02: The Collectors

You don't start out to be a collector. But at some point you take a look at all the plants that you have compulsively bought and staged, and if you know all their Latin names and requirements, and if you have at least a mental list of the ones you still want...you're a collector.

There's no shame in being a collector. And succulents aren't frog figurines or Victorian dolls (I apologize if those are your things). You are collecting living art. These plants are cool.

At the nursery, I observe the evolution. Someone will buy a plant that fascinates them. Then I see them off-and-on over a period of time, usually following some sort of plant theme. They might ask if I have the plant with the green wavy things that look like brains, and I'll point out one of the crested euphorbias. Later they come in asking for a *Myrtillocactus geometrizans*, cristate form. Now they're speaking Latin. The change has come. Time to join a cactus club!

A while back I made a delivery to one of my semi-regular customers, Bob. He had a typical succulent landscape in the front, but as I made it around back, I was struck by the shelves and benches full of plants, large potted specimens crowding the patio, and hanging orchid cacti under the trees. I saw the awesome fruits of a true nature lover's obsession, but another person might think there was something to be concerned with here—a plant hoarder or obsessive collector. I might think that too, if my backyard was not so similar. Yeah, he had a problem. His pachypodium collection was incomplete.

Bob was sensitive about his yard. I'm sure he'd seen the raised eyebrows before. He started to make the usual excuses: "Well, some of these I just offset myself to give away, and I've been meaning to thin it out..."

I cut him off with a hand on his shoulder, and told him, "Bob, it's OK. You're one of us". He laughed at the sympathetic gesture and we got on with the business of discussing his collection. As long as your hobby isn't taking food off the family table, well, there are worse ways to spend your time and money. Compared with orchid collecting, succulent collectors can take a vacation and not worry every day they're gone. Some of the plants might even appreciate being left alone for a while.

Euphorbia gorgonis

Haworthia truncata 'Lime Green'

Ariocarpus retusus

Ferocactus glaucescens v. Inermis

Plants on these pages were photographed at the annual San Diego Cactus and Succulent Society show. Plant shows are a wonderful way to learn about your hobby from fellow enthusiasts, and to see some amazing rarities expertly staged. Some are absolute museum pieces.

Titanopsis calcarea

A cactus collector's nicely potted and staged specimen plants.

Whether you've crossed over into the collector category or not, collecting natural things, living or dead, has a grand tradition. Eighteenth and nineteenth century Europeans, particularly the English and Germans, pioneered some of the first collections of the natural world during the great Age of Exploration. Rothschild used his status as a rich dilettante to send his minions around the globe to stock his obsessive personal collection of stuffed birds, pressed plants, gems, minerals, shells or whatever he could move onto next. These first-ever amassed collections of biology from the far corners of the earth became the first (private) museums.

The collector bug may border on the obsessive-compulsive at times. In Susan Orlean's novel, *The Orchid Thief*, the character John LaRoche strikes a familiar chord with many of us. He starts the day with a goal of acquisition, or to open a door to a new world of wonders. Most succulent fanatics have at least a bit of this nature. You see something different, track that specimen down and buy it, then notice a similar type, do some research, get several more, develop a list of those you still covet, and off you go (hopefully not off the deep end). Eventually you slow down and manage what you have, but there will always be something beckoning. Once your collection is large enough, hopefully your spouse won't notice the new addition hiding in the back.

One of my favorite collectors, and a truly inspiring individual, I know only as "Geoff from New York." Geoff has been blind since birth. I have always considered plants to have a mostly visual appeal, and having the benefit of sight, well, let's just say I was shortsighted. Geoff has a network of friends on the West Coast, and they have visited my nursery so he can indulge in our shared hobby. He can identify many plants by touch. Even cacti! In his younger years, he did the "reptile thing," just as I had. He moved on to succulents, which he still keeps in New York (and keeping succulents in New York has its own set of challenges). He also uses smell more than the rest of us to help identify plants. I was astounded at a few of the plants he identified at the nursery. I have a feeling he can tell when a plant needs watering or feeding better than I can. I hope I always keep the gift of sight, but if I ever lose it, I hope I can borrow some of Geoff's inspiration.

A collector's cadre of curiosities. It's amazing how fast a new greenhouse or set of shelves can fill up. You can always thin the collection. Or get more shelves.

ABOVE Group planting in a pumice-rock planter

TOP RIGHT Crested specimens such as this *Euphorbia enopla* crest take time to convolute this nicely and can be difficult to pry from a reluctant seller.

RIGHT This is an example of spiral-torsion cresting in a *Pedilanthus marcocarpus*. This plant was dubbed "the extrusion" for obvious reasons.

TOP RIGHT *Dudleya pachyphyum* is endemic to a few islands off the coast of Baja, and is still rare in cultivation. It is an elusive plant that you're not likely to find at your local nursery, not because it is hard to grow, but because it is slow to offset and difficult to start from seed. The best source for plants of this nature can be your local cactus club, or backyard growers or specialty nurseries that may sell online.

BOTTOM RIGHT *Boophane disticha* is a highly sought bulb succulent from South Africa. It awakens from dormancy with a show of wavy distichous (straight-line) leaves, and also has a huge red flower. This is a fun plant for pronunciation. Although it looks like it is pronounced 'Boo-fain', the proper pronunciation is generally considered more along the lines of 'Bo-Ah-fu-nee'.

TOP LEFT A show plant specimen of *Euphorbia suzannae* has taken on the guise of a two-headed chlorophyll creature. It looks friendly.

BOTTOM LEFT One of the Baja rock figs, *Ficus palmeri*, presented as a root-over-rock bonsai. These figs from dry climates cross over into the semi-succulent realm. See the chapter on succulent bonsai for more inspiration.

dudleyas contrasted with red bromeliads and aloes

Some succulent collectors stick to a preferred type, whether by genus, or more likely general appearance. When a particular category is somewhat satisfied, another may open up. A cactus collector may move on to crested plants, or perhaps medusoid euphorbias or caudiciforms. One of the benefits of keeping such a growing collection is the general uniformity of care. Benches of assorted succulents as seen on these pages can be watered on the same schedule, with the same light requirements, etc. There will always be a few, such as pachypodiums, adeniums, and lithops, that will have some special needs, in particular protection from winter rains.

Seeing a collectors' passion on display can make you appreciate both the work they have done in amassing and staging their living art, as well as perhaps the larger picture, which is appreciation of the fantastic varieties of plant life here on earth. Some of these plants seem to be beyond what the imagination could conjure up, as if from a different planet (and that might be planet Africa, as that is where many of the most amazing succulents come from). Thanks to collectors, some formerly very rare plants are now more available due to the sometimes laborious but rewarding process of successful pollination and propagation.

Kingdom: Plantae

Phylum: Tracheophyla

Class: Liliopsida

Order: Liliales

Family: Agavaceae

Genus: Agave

Species: avellanidens

Agave avellanidens

Interlude: *Agaves*

Agaves

Agaves are sometimes referred to as century plants, owing to the misplaced notion that it takes a century for them to reach maturity and bloom. They are monocarpic, meaning that they bloom once at the end of their life, but it typically takes anywhere from five to more likely 15 to 30 years for most to reach their final blaze of dying glory. By the time you notice the flower spikes' sudden appearance, the plant is hopelessly on its final legs, and the flower, which at first looks more like a giant asparagus spear, must be one of the fastest growing in the plant kingdom.

The appeal of agaves lies more in their dynamic, often imposing structure, and their sometimes extraordinary color schemes—from various shades of blue or green, to all manner of variegation, and usually vicious spination. Agaves and aloes have a similar look, but while only a few aloes will draw blood, there are only a few agaves that won't make you bleed. Many get quite large and are very drought-tolerant once established. The larger agaves are good candidates for wide areas of land or slopes that are difficult to water. Once established, they will take up residence for many years, the offspring taking over for the flowered-out adults. Mid-sized to small agaves can make excellent container specimens or garden accents. You can snip the spine tips if they poke you one time too many.

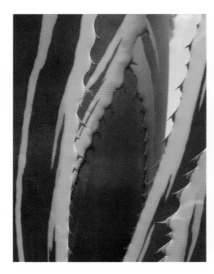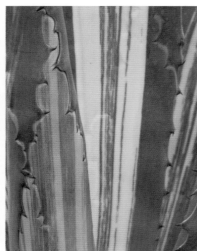

Agave guadalajarana (top) *Agave parryi v. truncata* (bottom)

TOP PHOTO *Agave guiengola*   **BOTTOM PHOTO** *Agave parviflora*

There are hundreds of agaves to choose from, with an equal number of hybrids. Below are some of the more frequently available and/or desirable agaves for a Mediterranean climate, segregated by size.

Large agaves: (4' x 4' to 8' x 8') *Agave americana* (blue or variegated), *Agave franzosinii*, *Agave ferox*, *Agave weberi*, *Agave tequilana*, and *Agave ovatifolia*. There are also several large furcraeas (see pages 88/89) that might as well be agaves. The variegated forms of *Furcraea foetida* and *F. selloa* are particularly attractive.

Medium agaves: (2' x 2' to 4' x 4') *Agave* 'Blue Glow', *Agave* 'Red Margin', *Agave* 'Blue Flame', *Agave attenuata*, *Agave shawii*, *Agave colorata*, *Agave potatorum*, *Agave gypsophila*, *Agave bovicornuta*, *Agave desmettiana*, *Agave parryi* (various forms).

Small agaves: (max size less than 2' x 2') *Agave victoriae-reginae*, *Agave ocahui*, *Agave schidigera*, *Agave Keiji-jo-kan*, *Agave applanata*, *Agave parviflora*, *Agave macroacantha*, *Agave* 'Royal Spine', *Agave* 'Dragon's Toes', *Agave lophantha* 'Quadricolor'.

Agave bovicornuta

Agave potatorum v. verschaffeltii

Agave ovatifolia

Agave lophantha, quadricolor variegate

 As with many succulents, agaves can be starkly different in appearance, yet somehow have just enough sameness that you can identify one as being part of the group. With few exceptions, they have sharp and rigid spines, and relatively inflexible leaves as compared to aloes. The main exception is *Agave attenuata*, featured on the next page (82). They can range from monsters like *A. americana*, *A. franzosinii*, *A. ferox*, or *A. guiengola*, to midsize 3'x 3', such as *A. bovicornuta*, *A. parryi*, or *A.* 'Blue Flame', to dwarf varieties like *A. parviflora*, *A. lophantha*, or *A. ocahui*. They range from frost-tender subtropical parts of Mexico to snow-tolerant plants from the Rocky Mountains.

The annual blossoming of the "elephant trunks" in Mediterranean regions calls attention to clumps of *Agave attenuata*, also known as the "foxtail agave" to use another animal metaphor. *A. attenuata* is one of the few soft, non-spiny agaves, very distinct from its kin. They can be prolific clumpers, and can either be a focal-point plant as seen above, or better yet used as a soft background border to a succulent garden. Less often seen is a bluish cultivar known as *A.* 'Boutin's Blue' (inset above, which may in fact turn out to be a distinct species) and various stunning variegated forms, one example of which can be seen on page 193.

Agave americana medio-picta alba

This agave flower must have acquired a virus that contorted its branches into a cartoonish shape.

An *Agave americana* blooming beachside at Las Gaviotas, Baja, CA, Mexico.

Consider the terminal agave flower as a plant going out in a blaze of glory at the end of days, giving itself its own magnificent funeral pyre. After a month or two of growing higher and higher, the flowers slowly begin to display from mid-stalk to top, attracting all manner of bees and birds. Often fertile seedpods are left for people who are willing to drop the stalk and make the effort to germinate them (seed germination occurs in natural habitat, but rarely in cultivation). The dead flower stalk can stand for years as an artistic sentinel, a dried tree serving as the gravestone of its former self. And because most agaves will have offset at the base prior to blooming, usually new ones remain to take their place. Left untended, *Agave americana* will form an impenetrable thicket after many years. Most clumping agaves will form suckers or pups that can be culled for new plants, but they do that fairly early in their lives and then go through "agave menopause" and don't produce any more. Then seed collection is the only hope to recreate your beloved plant. Some agaves skip the seed phase and instead produce bulbils along the stalk, which eventually fall and can be collected as ready-to-go plantlets.

Agave victoriae-reginae compacta, beginning of the end. In a matter of weeks the flower will grow to 8 feet high, bloom over a period of months, and then the plant will slowly die. Agaves live long and die spectacularly. Appreciate it for the event that it is. The hummingbirds and bees do.

Agave ferdinandi-regis

Agave macrocacantha

Agave pygmae v. 'Dragon's Toes'

Agave pumila

Agave ocahui

Agave zebra

Plants that die when they flower are called "monocarpic." Look at the size of the bloom relative to the *Agave zebra* flowering in the photo (bottom right). That little two-to-three foot rosette likely showed its first sign of blooming a little more than a month before this photo. It seems to have doubled in size during that time. No wonder it will not survive.

Agave *Keiji-jo-kan,* variegated

Agave 'Royal Spine'

Agave 'Gentry Jaws'

Agave *schidigera,* variegated

Agaves are generally not pest-prone plants, but there are a few concerns. There is an agave weevil that can take down a fairly big plant; look for splitting in the leaves or a general droopy appearance. There is also a new form of agave mite that is becoming a menace. Unfortunately, expensive systemic poisons are the only cure at the moment.

Many of the agaves on these pages, particularly the variegated forms, did not exist in cultivation even a few years ago. Those that did were very rare and could command hundreds of dollars for a two-inch pup. While that is still the case for some, in general the ease of propagating agaves has led to more plants available to the public at perhaps still premium, but at least somewhat lower, prices. And new varieties continue to be introduced. *Agave* 'Dragon's Toes', a form of *A. pygmae,* is one example.

Why am I getting spots on my plant's leaves?

First, consider that plants in habitat are almost always somewhat marked up and stressed; such is the life of a succulent. Leaf spotting can be caused by lots of things, and is rarely anything you did wrong. Hail can leave impact craters. Snails can munch holes in tender succulents, and even chew on tougher agave leaves. I've had agave leaves burn a bit during a sudden spring heat wave; it just got too hot too fast. There are a number of environmental factors that can cause isolated tissue stress. One myth is that overhead watering can cause plants to burn; perhaps that is the case with some tender tropicals, but I've seldom if ever noticed that with succulents. Although a damaged leaf won't ever repair itself, take consolation in the fact that new leaves will slowly push up and old leaves eventually will shrivel away. It takes a year or two for the hail damage to totally disappear via new growth, but it gets there. Learn to live with imperfect plants.

Agave franzosinii, Rancho Soledad form

Furcraea foetida (gigantea) medio picta is a formerly rare plant that has become less so in recent years via tissue culture (laboratory cloning) and offsetting. The variegated form has eclipsed the green version, which is almost extinct in cultivation. It is a striking sentry plant, either individually or as a background grouping, as seen above. In the foreground is a mass planting of *Agave* 'Blue Glow', along with blue *Senecio mandraliscae* and *Euphorbia milii* 'Dwarf Apache'.

A note about *Agave* 'Blue Glow' seen above and elsewhere in this book. There are several agave hybrids with names like 'Blue Glow' or 'Blue Moon' or 'Moonshine', but the true 'Blue Glow' in this author's opinion is *A. attenuata x A. ocahui*. It has a stiffer leaf and slightly more blue color. The other one frequently seen is *A. attenuata x A. shawii*, which was first called 'Red Margin', seen on the cover image of this book.

It is a little greener and has a softer leaf, more like *A. attenuata*. And to further complicate matters, another hybrid known as *Agave* 'Blue Flame' (seen in the variegated form on page 195) is a hybrid of *Agave attenuata* and *Agave celsii*. A bit confusing, but the result is we get more plants to play with and argue about.

Flower stalk of the
Dasylirion longissimum

Yucca 'Blue Boy'

Dasylirion wheeleri

Furcraea foetida medio-picta

Dasylirion longissimum, old plant

Yucca rostrata

Yucca gloriosa (elephantipes)

YUCCAS, DASYLIRIONS, FURCRAEAS

Yuccas and dasylirions are also part of the agave family. Unlike agaves, most are not monocarpic (they will live to flower again). Most are true desert plants, and are very popular in architectural desert landscapes. Highly prized *Yucca rostrata* and *Yucca rigida* are blue beauties with large creamy-white flower stalks. The most famous yucca is likely *Yucca brevifolia*, the "Joshua tree" of the California high deserts. That's where they prefer to live, although they may survive closer to the coast. Most dasylirions don't have a trunk as many yuccas do. Dasylirions make dramatic statement plants in any succulent or cactus garden. Furcraeas most closely resemble agaves, but are generally softer-leaved and less toothy. *Furcraea foetida medio-picta* is a spectacular, bright-yellow-and-green striped specimen that commands attention in any garden. Another related plant that adds nice flowers to a desert landscape is *Hesperaloe parviflora* (not pictured).

CHAPTER 03: The Growers

Anyone who has accumulated at least a small collection of succulents has at some time become an accidental propagator. If you own a jade, a piece will fall off, you lay it in a pot, and up grows another plant. Not all succulents are so easy to start, but many can be grown by detaching a rosette (sometimes with new roots already hanging off), or snipping a cutting, or perhaps planting a detached leaf and seeing new growth cluster at its base.

You have to admire succulents' ability to survive. I have had enthusiastic customers over the years who bought many plants in binges, but over time I wouldn't see them as often, because their yard or collection became full. Most have figured out how to propagate more plants from what they already have, and will even start to bring me back some of my original material to trade. Some joke that they're thinking about starting their own nursery. I hope they're kidding. Luckily, there is always a plant they don't have yet that I can tempt them with.

If you are paying what seems like a lot for a small plant, it may be a type that can't be propagated as discussed above (vegetative propagation) and must be grown from seed, such as many types of cacti, certain aloes, and most rare caudiciforms. Growing from seed can be a rewarding challenge, but it takes a commercial grower with a nice greenhouse setup to really get a volume of these plants into the market.

Northern San Diego County and Tucson, Arizona are among the leading regions for wholesale cactus and succulent growers. Some grow rarities in greenhouses, others excel at "chop and plant" in rows in outdoor fields. Sun-grown plants already under the duress of the elements make a smoother transition to your home. Landscapers are becoming more savvy to the world of succulents, with impressive specimens making their way into commercial and residential landscapes. A succulent landscape used to announce the house of a plant nut; that isn't always the case anymore.

Commercial cactus (years ago "cactus" was the catch-all phrase rather than the more appropriate "succulents") propagation got its start in the United States primarily in the Los Angeles area and portions of Arizona in the early part of the 20th century. Harry Johnson ran one of the first destination cactus nurseries in Paramount, California. Others followed, many of them eventually moving 100 miles south to the benign climate and rural hills of Vista, California. Many have come and gone, yet a few old-timers remain, including several large operations that ship millions of succulents across the United States and overseas.

ABOVE The house of a collector and backyard grower. If you have enough stock of rare and mature plants, why not collect seed or take cuttings to help support your plant addiction? Backyard growers are often dedicated to propagating some of the more unusual collector plants that may not have as much commercial zing as the colorful aeoniums, echeverias, etc. They trade amongst themselves, sell via the Internet, or supply to local nurseries. Long live the little guy!

FAR LEFT A grower shows off his mother plant of Echeveria cante, relatively new to the trade and a show-stopper beauty of a plant. They are still fairly rare and command a high price, but as you see here, they are being propagated—still by seed as they rarely offset. As with dudleyas, don't touch the leaves; you will leave fingerprints in the fine powder.

AT LEFT Hand-watering with a rain nozzle is the most effective method of coverage. This worker seems unhappy with his meager wages, but still prefers watering to pulling weeds.

    We retailers rely on medium and large growers for our inventory, but one of the fun parts of the business is dealing with the backyard grower. These inveterate plant people are usually collectors (see the Collectors chapter) who either are following their own internal imperative to propagate more plants, or have figured out a way to pay for their hobby by making more plants out of those they already have. Indeed, some of the rarities and oddities aren't economically worth the effort for large growers, and are only available via the backyard grower. Because there is always something new and different in the plant world, we will often do a bit of horse-trading to get something new into the hands of someone who is good at making plants into more plants. For example, the only way to make new crested plants is to cut up a large, mature specimen into several smaller ones. Those of us who worship and know the value of these plants just can't do it. But the mad propagator just sees more plants. Hack away, my good man.

Fields of fire. Here is a cutting field of Euphorbia tirucalli 'Sticks on Fire' at Western Cactus in Vista, CA. Repeated cutting makes for more new growth, hence even more vibrant color, especially in winter months.

Featured here are a few pages from the Johnson Cactus Gardens catalog, circa 1940. Before the age of Google Images, this was the source of dreams and inspiration for cactophiles everywhere. Color photography was expensive, so there were many hand-painted renderings. Customers sent in pictures of themselves posing with their cactus collections, and Johnson would put the photo of Mrs. Merrill Gutzmer of Whitewater, Wisconsin in the catalog. Little treasures were shipped across the country, and I bet a few, or perhaps their progeny, are still alive today. And only 19 cents for a peanut cactus! But the 3% California sales tax might have been a sore spot for some.

Can I take my plants with me when I leave the state?

I'm not going to attempt to address any legal issues here for obvious reasons, but I can supply a few generalities. Agricultural states such as California, Florida and Hawaii are more concerned with potential invasive plants, and in particular the pests that might hitch a ride on them, than are states that have freezing winters that will kill off any unwanted interlopers. Some border crossings will allow you to take the plant but not the soil attached, so bare-root plants stand a better chance of relocation than those in soil. The plants may have to be sprayed as well. Succulents are fine either shipped or carried in a bag bare-root, providing they don't freeze. There really is no reason to carry all that heavy dirt with you anyway. If you have established succulents in the ground, consider taking pups or cuttings with you if you have to move, although large succulents do come out of the soil fairly easily because most are shallow-rooted.

Do you have that little cactus that looks like a little ball thingy?

In this case I think I know what you are talking about, probably *Euphorbia obesa*, also known as the "baseball plant" (to me they look like sea urchin skeletons), see page 185. Sometimes the request is less descriptive, such as "kinda green with some leaves and branches, sort of little spines." That narrows it down from thousands to hundreds of possibilities, but if you really are looking for a particular plant, those of us in the business are fond of Linnaean binomial nomenclature. Genus and species when possible, please. That way we know we're talking about the same thing. Often a plant is identified not only by genus and species, but also by its cultivar name, such as *Echeveria agavoides* 'Lipstick'. *Aeonium* 'Sunburst' usually goes without a species name, as its parentage information has been lost over the years and can only be speculated. And to further confuse things, botanists periodically reclassify plants into new categories. *Abromeitiella brevifolia* is now *Deuterocohnia brevifolia*. Naming protocol changes as well. *Aloe bainesii* is

now officially *Aloe barberae*, due to research on who was the first to submit a description of the plant to the Royal Botanical Society in the 1800s. We stodgy old-timers have a tough time unlearning names, however. It's still *Aloe bainesii* to me. Common names are sometimes quite appropriate and sometimes funny, and often are well enough known to avoid confusion with anything else. "Mother of millions" will almost always be one of the kalanchoes that forms many baby plantlets on the leaf tips, but there are several different kinds in cultivation. The Dr. Seuss tree is probably *Aloe barberae*, but might be a large *Kalanchoe beharensis*. The "mother-in-law's chair" must be some kind of nasty cactus, but I'm never sure which one. That might be up to the discretion of the son-in-law. The jade crassula is one of many plants referred to as the "money plant." In the nursery business, the money plant is whatever is selling hot at the moment.

A photograph is probably the best way to find the plant you're looking for. Most nurseries or growers will know right away what it is. Beware of some of the exotic rarities you may see in magazines or the Internet. Even those well-connected in the plant world can have a tough time locating many plants, and there are plants that look wonderful in habitat photos that have just never successfully made it into cultivation. Some plants are easy to grow but very difficult to propagate. If you are so inclined, you should keep the labels of the plants you have that were properly identified when purchased, and try to identify the rest of your collection. It's good to know what you have, and you can impress your friends with your knowledge of Latin. But again, we're talking about succulents. You can remain blissfully unaware or quasi-clueless and still manage to keep them alive and happy.

Beware all who dare to enter. This is a spooky forest of *Euphorbia ingens* planted as four footers for later cutting into new stock. Supply can grow faster than demand, and now this grower has one of the more impressive stands of *E. ingens* outside Africa. It only took 12 years! Euphorbias grow much faster than their cactus cousins.

When visiting wholesale growers, after a while you learn that there are plants they just won't sell. Don't bother asking. Growers are collectors at heart, and may have plants that are one-of-a-kind, or that they have become attached to. If growers collect their own seed, they need to keep their parent plants together for annual pollination and seed harvest. Dioscoreas are propagated only by seed, so it is in everybody's best interest if a disciplined grower has his male and female plants together as a couple, with either natural living pollinators close by or a paint brush ready to do the job.

Starting plants from seed is a combination of art, science, and a little luck. A ventilated greenhouse is ideal, but most of us don't have that luxury. If you obtain viable seed, it is worth a try at home. You can make a miniature windowsill hothouse with a 6- or 8-inch clear plastic container.

Sprinkle seed (most cactus/succulent seeds are small, sometimes almost dust-like) on a fairly inert medium such as washed concrete sand that contains a minimum of organic material. Cover with at most a trace of sand. Place your small pot (plastic or Styrofoam cups are OK) inside a larger plastic container, wet thoroughly, cover with the lid, place in bright indirect light, and hopefully in a few days or weeks, you'll see green popping up. Remove the lid to ventilate, and continue wetting and covering as you see fit.

The hardest part isn't germination, it is keeping the little ones going. The trick is to keep the seedlings fairly moist, but not allowing moss or fungus to take over the environment. That is where ventilation helps, or you can experiment with a fungicide. If you end up with three mature plants from 20 seedlings, you've done well. I've enjoyed some amateur success with growing aloes from seed, but because I can't control my outdoor pollination, I often end up with a mixed bag of true species and hybrids. Sometimes I like the hybrids more than the true species.

A grower's stock colony of *Dioscorea elephantipes*, with male and female plants for annual seed production.

Do you go out to the desert to get your plants?

No, that is generally illegal and unethical, and mostly unnecessary as most succulents are from Africa, or Central and South America.

Do you travel to Africa to get your plants?

I wish. I have yet to see most of my favorite plants in habitat, except for those from Baja California, Mexico. The African, Old World starter material has been introduced slowly over the past 150 years and is propagated locally, so field collection for commercial purposes is now restricted to experts who find the increasingly rare new material for slow introduction to the trade via botanical garden introductions. Also, many plants don't occur in nature, but are engineered hybrids of plants that would never have met on their own, or unusual "sports" that originated in cultivation. I'll bet almost a third of the inventory at my nursery today are plants that were either unavailable, undiscovered or yet-to-be-engineered when I started my nursery twenty years ago.

Photo by Bob Wigand

Colors and textures. If you squint, this can almost look like a series of watercolor strokes across the page. In front is *Aeonium* 'Zwartkop', above is *Aeonium arboreum* variegated, then a swath of "old man" cactus, possibly *Oreocereus celsianus*, with a topping stripe of aloes, likely *Aloe africana* or hybrids of the same. The far background is the native chaparral of inland north San Diego County, which has an ideal climate for growing these African, Canary Island, and South American plants. As real estate values rise, much former greenhouse space or outdoor growing fields are replaced by housing tracts. One advantage of a slow economy is the preservation of native hills and low-water, small-footprint agricultural pursuits. As stated elsewhere in this book, these non-natives are good about staying put, and rarely intrude into the natural habitat.

Growers will on occasion graft plants, which involves removing the roots from the display plant and attaching it to a more vigorous stock. This can be done to encourage a difficult plant to grow better, or as in the case with the "hothead" *Gymnocalyciums* above left, it facilitates and encourages a color abnormality that is usually minimally apparent when it grows on its own roots. Grafted plants have their own appeal, but there is something sort of "Frankensteiny" about them; the head of one plant on the body of another. Often over time the head, or "scion," can emaciate away, and the body, or "stock" can begin to grow in its place.

ABOVE LEFT This photo of the above-mentioned 'hothead' cacti was taken in a greenhouse. The extra heat can really force a plant into color, but may be difficult to replicate in a home environment.

ABOVE RIGHT A flat of *Crassula* 'Morgan's Beauty', also known as *C*. 'Morgan's Pink'. It is one of many hybrids of *Crassula falcata*, and if it always looked as good as these, would be recommended as a must-have plant. Unfortunately, it is prone to a type of fungus known as "rust," which is an apt name, as the leaves appear to get rusted over in time. It is one of the few succulents that you buy knowing that you may only have a short time to enjoy it, but priced at $3 to $4 for a small one, what the heck?

AT RIGHT A field stock of *Portulacaria afra variegata* at Waterwise Botanicals, Bonsall, CA.

A grower's stock of crested *Trichocereus pachanoi*, showing some reversions to regular columnar form.

Kingdom: Plantae

Phylum: Magnoliophyta

Class: Magnoliopsida

Order: Caryophyllales

***Family:* Cactaceae**

Genus: Echinocactus

Species: grusonii

Mammillaria spinosissima cv. 'Un Pico'

Cacti

Cacti are an acquired taste. Many are little generic spiny things that don't have a lot of character, other than an occasional and surprising bloom. Others are dynamic sculptures. When I first got into the business, like most of my customers, I had a stronger attraction to the non-cactus succulents. I was advised by some old-timers that my view would change, and it has. As with most succulents, there can be some very subtle and sublime design elements that slowly become apparent with cacti, punctuated occasionally by a brief but startling display of flowers. When cacti want to attract pollinators, they pull out all the stops.

The cactus family is a huge group of New World succulents, with a common morphology of fat bodies (spherical, columnar, or in the case of opuntia, large pads), almost always protected by impressive spines.

Is the plural of cactus "cacti"?

Yes, it sounds better than "cactuses." But in common speech, the phrase "Wow, look at all that cactus," referring to a group of cacti, sure sounds plural enough to my ears. In fact in a few spots while writing this book I had to go back and correct several "cactus" to "cacti," including that last sentence right after I wrote it. I certainly never make it a point to correct someone using the word incorrectly. Unless, of course, they are calling non-cactus succulents "cactus" (or "cacti"). I hate to let a teachable moment pass by.

Some African euphorbias have a very "cactusy" appearance, but there are a couple of ways to tell the difference. Euphorbia spines can be sharp, but are much less likely to detach and stay in your fingers. Cacti like to leave a little part of themselves with you. In the case of the "jumping chollas," they leave entire segments attached to offenders, to be painfully carried about until they eventually detach, fall and start a new plant. Some cacti have irritating glochids, fuzzy hair-like spines, that stay in you like fiberglass; less painful but more irritating. Perhaps the biggest difference between euphorbias and cacti are the flowers. There is no contest here; cacti have some of the most beautiful flowers in the plant kingdom. Hybridizers have been working for years to develop new and more amazing flowers, particularly with echinopsis, trichocereus, lobivias, and epiphyllums (orchid cactus).

Cleistocactus strausii

Echinofossulocactus

Pachycereus pringlei

Opuntia sulfurea

Rhipsalis baccifera

Cleistocactus strausii

Pseudopilosocereus azureus

Ferocactus latispinus

Epiphyllum flowers

Like all succulents, cacti can be kept as container plants or grown in-ground as part of your landscape. Collectors favor exotics such as ariocarpus, rebutias, small mammillarias, echinopsis, or copiapoas for specimen container plants. For landscapes in appropriate climates, probably the best member of the cactus family to consider is the golden barrel, *Echinocactus grusonii*. Also popular are the various iterations of blooming trichocereus or trichocereus/lobivia hybrids. Columnars to consider include *Pachycereus pringlei*, the silver torch *Cleistocactus strausii*, or the various blue forms of pilosocereus. For larger background "hedge" cacti to frame a garden, consider the prickly pear opuntias, particularly the pink/purple hued forms of *Opuntia* 'Santa Rita'.

Sometimes you have to a see a family or genus of plants in their collective array of shapes and sizes and flowers to truly appreciate them. Some people have to look past the spines to appreciate cacti. Others learn to appreciate them for their spines. The *Ferocactus chrysacanthus* (facing page) is truly impressive in both the density and bright coloration of its armaments. "Go ahead and eat me. I dare ya."

Tephrocactus articulatus

(Or really more of a statement): I have grandkids and don't want anything spiny that might poke them!

Well, there really is no proper response to that. The customer is always right. Most cacti do have some nasty spines (although I encourage them to touch the "spines" of most aloes and many euphorbias; they can sometimes be menacing only in the visual sense). Spination is an effective evolutionary trick to stay uneaten, or maybe unloved by a cautious owner. Try not to let the spines keep you from enjoying the plant. Many people have a built-in prejudice against cacti, usually due to some bad experience in their past. Even if you are an anti-spinite, try a few strategically placed golden barrels in a succulent land-scape. They really add drama and architecture.

Ferocactus gracilis   **FACING PAGE** *Ferocactus chrysacanthus*

A grower's field of golden barrels (*Echinocactus grusonii*)

An unusual astrophytum

In garden settings, few plants stand out like golden barrels; they look like large, glowing, golden orbs when backlit. Columnars such as cardons, cleistocactus, and especially blue pilosocereus provide a nice slice of architecture either in-ground or in large pots. Other miniature cacti are perfect for a patio or sunny garden window. Small mammillarias, notocactus, parodias—the list can go on and on—will live for years between repottings. Cacti are among the most drought-tolerant of all succulents, but they still appreciate a regular drink and occasional shot of plant food. Provided they are planted in a cactus mix, don't hesitate to water when dry. Many are from summer-rain habitats.

FACING PAGE *Mammillaria laui ssp. subducta*

Is there anything more contradictory, more oxymoronic, more yin vs. yang, than soft pastel papery flowers next to sharp and vicious spines on the same plant? Cacti have figured out how to work them both together. "Pollinators, come hither. Everyone else, you wanna piece of this?"

Ariocarpus trigonus

Mammilaria geminispina, right is *Mammilaria spinosissima* 'Un Pica'

Specimen cacti

Unidentified cactus, likely a minimal spined version of *Stenocereus stellatus* or a trichocereus

If you have the good fortune to attend a cactus and succulent show, take your time at the cactus tables. The other succulents usually demand more attention for their crazy shapes and dizzying colors, but there is a more refined elegance in properly grown and staged cacti. Most shows are in spring or early summer, so you will see some impressive flowers on many of them. Others, such as the *Ariocarpus trigonus* (top row left), display a unique adaptation of "hair" to protect both the crown of the plant from sun as well as the new fruit until it is ready to go. This cephalum can grow to be taller than the plant itself, and is especially prominent in *Melocactus*. Bottom left are specimen cacti on display at a cactus and succulent show. These clumpers are rebutias and sulcorebutias. It can take many years for plants to become such magnificent clusters. Bottom right is an unidentified cactus, likely a minimal-spined version of *Stenocereus stellatus* or perhaps a trichocereus, that looks like a cluster of green corn cobs.

Copiapoa haseltoniana

Astrophytum asterias

Gymnocalycium baldianum

Ferocactus palmeri

Ferocactus chrysacanthus

Ferocactus latispinus

Do you wear welders' gloves when you're handling a cactus?

No, I don't like gloves because they can give you a false sense of security. Many cactus spines will go through most gloves. I've spent years planting large golden barrels in the landscape, and I just dedicate a blanket or old beach towel as my cactus cradle. When you de-pot a cactus, shake out the dirt and it's just some limp roots and all body. I use a towel as a sling and just drag the plant into place, barely digging a slight depression to plant in. I use my boot to gently nudge it into place, kick some dirt around the perimeter, and the job is done. An important tip from firsthand experience: never use it as a towel again!

What do you do about the spines when you do get stuck?

In my experience, it is usually the plant next to the one you are working with that gets you. You feel it get you, it hurts for a minute, and then it's not so bad and you think maybe it just went in and out and didn't leave a tip in you. Sometimes that is the case. However, if it still hurts the next day, you likely still have a little piece of spine in you. I'm convinced some have a mild toxin or irritant that makes it a little worse. You should get a sterile needle, break the skin, and if you can find the foreign object, scrape back and forth until it pops out. I've had a few that I never got out, and I assume my body absorbed them after a few weeks of itching. I guess I'm slowly going to transform into Cactus Man. For glochid spines, some folks prefer tape or let some white glue dry on their skin and then peel it off, hoping the spines come with it. Scraping back and forth with a credit card sometimes works as well. As a last resort, go see your doctor.

Opuntia microdasys, a beautiful "glowing" cactus. However, the warm fuzzy glow coat is actually glochid spines, little clusters that stick in you like a bad fiberglass rash. They don't go in deep, but if you brush into one, they come off in numbers. It is a lovely cactus, but handle with caution!

Cereus peruvianus, a fairly ubiquitous columnar cactus in California. It may reach such a massive size, it can dwarf a small home. Its exact origin is supposed to be Peru, although a locality for the original population has never been established. At least it is abundant in cultivation, offering huge clusters of overnight white flowers in warm months. Flowers last through a sunny day, differentiating it from several other cactus claiming to be "night-blooming cereus," whose large flowers wilt in the first rays of bright sunlight.

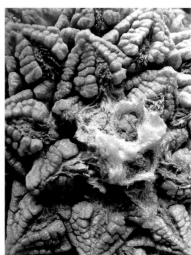

One of the collector cacti, *Ariocarpus fissuratus*. All ariocarpus are very slow growers, and posses a spineless architecture that doesn't immediately suggest "cactus." They do have beautiful flowers in late summer and fall.

Ferocactus glaucescens, a beautiful blue cactus, accentuated in this rare specimen by the lack of spines. The yellow flowers are nice, but the pre-flower buds can be equally impressive.

Stenocereus marginatus

An unidentified opuntia, likely *Opuntia polyacantha*, blends nicely with sage and senecios. There are a large number of opuntias, some spiny, some not, some in between. Many get very big, so plant accordingly. They can have showy flower displays spring into summer, followed by edible fruit.

Mammilaria geminispina

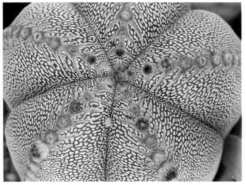

Astrophytum asterias 'Super Kabuto', a Japanese cultivar

Parodia magnifica

The Christmas cactus, a hybrid form of a *Schlumbergera*. These semi- tropical "orchid cactus" relatives make fine indoor plants provided they have some bright light. Often they bloom closer to Thanksgiving than Christmas, but older plants will put on a long bloom show. Treat a bit more like a tropical than a true cactus. There are several flower colors available.

Notocactus

Mammilaria theresae

Here are various forms of flowering cacti. On this page is a spineless form of *Tacinga inamoreana* with flowers that evolve from violet buds to orange flowers. Shown on the facing page are primarily hybrids of trichocereus, which show the varying colors that blaze into brief glory in the spring. There is usually a big initial spring flowering, followed by intermittent flowers throughout the summer and early fall.

FACING PAGE Flowering cacti photos by Bob Wigand

CHAPTER 04: Succulent Landscapes

If you are tired of your high water bill and don't really need all that thirsty lawn, and especially if you have an enthusiasm for plants, then you should get to work on a succulent landscape.

Installing the plants is the fun and easy part. Most of the hard labor is usually in killing and/or removing your lawn or previous landscape, adding mounds, and importing rocks and boulders. Some succulent gardens just sort of appear over time as plants get added to an existing landscape, but if you want a more pulled-together garden, there are a few loose principles you should consider.

Succulent gardens look best if they are built in the context of topography. You can do a modern or geometric garden using succulents, but they seem to display better in a less overtly man-made venue. If you somewhat mimic a desert arroyo, albeit using plants from different continents, you'll appreciate the plants' natural appeal. Rocks and elevations evoke a slice of the natural world and better highlight your plants.

I recommend temporarily removing existing succulents so you don't try to plant around them and compromise your design. Once you have cleared your planting area to a blank palette, build several mounds with some type of amended topsoil or sand/loam mixture. Good brown dirt is usually OK. Mounding gives you visual appeal as well as allowing your succulents to enjoy good drainage. Next move in some boulders. Bigger is better if you can afford it; at least a third of your budget should go towards soil and rocks. If you have access, a crane placement of several large boulders, with medium and smaller ones in proximity, is the best first step in garden design.

If you have a primary view of your garden, have the high points of the mounds towards the back, and work towards a lower plain with perhaps some sub-mounds in front. Usually a dry river bed will appear whether you plan for it or not, so identify it and leave that as a primarily unplanted area. That riverbed, along with a few large boulders, will force you to have some much needed blank spaces, because succulent gardens tend to fill up over time. It also provides access. Stand back and look at your plant-less rock garden. If you did a good job, it should have artistic appeal in a dry-desert sort of way. Take a picture. Now you get to do the fun part.

A Michael Buckner desert-style garden with stark cactus sentinels.

The succulent glade above is an established garden in Rancho Santa Fe, CA. Restraint has been exercised here, with repetitions of soft plants like *Agave attenuata* and an *Aeonium canariense* hybrid. At left is *Aloe plicatilis*. On the other side of the path, the garden transitions into a sunnier and drier area, so the plants transition as well, into more cacti and agaves, following a blue color scheme. Some of the plants above also merge into the dry side to suggest a natural blending as might occur in nature.

Furcraea selloa is a centerpiece in this colorful garden.
Photo by Randy Laurie

One design principle is the Rule of Odd Numbers. There is a tendency to either have one or three or five of a particular plant. I can sort of get behind this philosophy, as a group of three can have a pleasing effect. However, I've seen plants presented as a pair, and it looks great. And beyond three, I really don't think most people are counting, so don't get too hung up on the odd-numbered thing.

FACING PAGE

TOP LEFT The large blue agave is not the common *Agave americana*, but a form of the more powdery blue and slightly larger *Agave franzosinii*. This garden is mere steps from the sand at Moonlight Beach in Encinitas, CA, and is becoming a tourist attraction, especially when in full winter bloom as seen here. This garden is also featured on page 126.

BOTTOM LEFT A seaside succulent garden shows the usual enthusiasm of its owner for these living sculptures. Many succulents are winter bloomers, such as the aloes and aeoniums shown here. Cooler weather also can accentuate the leaf coloration in many plants as well, so December and January are usually some of the best months in the garden. Sadly, the winter rains also bring out the weeds. Try to get them before they set seed. Design by Rogue McNeal.

TOP RIGHT An African/Mediterranean dry river bed featuring aloes and xerophytics. Design by Jeremy Spath.

BOTTOM RIGHT A *Dasylirion wheeleri* snakes out a convoluted flower in a Leucadia, CA, succulent landscape.

Hopefully you've been collecting succulents in anticipation of planting your garden. Even if the project is months away, if you fall in love with a plant you see at a nursery, buy it. If you don't, it will likely be gone when you go back for it later. Succulents are happy biding their time in plastic pots. You may have noticed such a potted plant starting to radically outgrow its container, realizing when you try to pick it up that part of the reason for its sudden happiness is that it has rooted itself in place. A succulent can get impatient, and if you don't plant it, it'll do the job where it sits. I have several huge euphorbias at my nursery that have declared squatter's rights to the little corner I set them down at; just a few shards of imbedded plastic offering mute testimony to their former portable lives.

We plantaholics are never done adding plants. But there are steps you can take to at least pull your succulent jungle together enough that it doesn't look chaotic. Start with your icons, or larger plants that will be your main focal pieces. You may have a large tree aloe or euphorbia that lords over the garden, with several large aloes or columnar cacti. After you've set your major pieces (again, more in back than in front), in round two you begin repetition of mid-sized (say 5-gallon pot sized) plants. Here you can still have single individuals, but you want to begin introducing some repetition. Perhaps four to five golden barrels, say three clustered together, then two others scattered about. You can start small, but having some larger plants as well gives your garden a more finished look up front. Lastly, choose several smaller plants (ones that won't ultimately get big, or at least tall), and plant these in drifts or groupings. Repetition is important here. Choose at least one smallish plant as a sort of theme that winds through your garden. The yellow-orange *Sedum nussbaumerianum* or blue *Senecio serpens* serve this purpose well. I like to plant in small islands clustered around the boulders, as that is a look that appears in nature. Succulents are slow growers, but within a year or two they will begin to spread.

TOP LEFT Try to resist planting too densely at first; plants need room to grow. This tableau is in actuality a truck bed full of plants on the way to be planted in a new succulent garden. You can see how the colors and textures and varieties of shapes work together. Cacti, aloes, aeoniums, cotyledons and crassulas can all handle the same care in the landscape.

TOP RIGHT *Aloe dichotoma* is a wonderful piece of botanical engineering. Here it serves as an entryway sentinel in a Southwest-style garden with a sparse, desert appeal and fewer floriferous plants. This doesn't necessarily mean you are restricted to plants from the desert Southwest; the aloe here is from the Namibian desert of Africa. An international garden of like-minded specimens can inspire you to someday visit their relatives in their native habitats.

BOTTOM LEFT In addition to natural boulders, a flagstone path can be both an essential and aesthetic addition to your garden. Much like a dry river, it provides needed blank space, and a nice viewing platform into your planted mounds. The groundcover here is *Dymondia margaretae*, a drought-tolerant plant that can handle foot traffic.

BOTTOM RIGHT Broken and stacked flagstone creates a rustic planter. Pieces of demolished concrete hardscape are also acceptable, and can be stained in earth tones.

Gravel pathways provide access into, as well as a framing device for, the sometimes-crazy abundance of a succulent garden.

Landscape design by Frank Mitzel

ABOVE LEFT This *Aloe speciosa* offers a sculptural statement to a courtyard's aloe-themed garden.

ABOVE RIGHT Combine full, bushy succulents for a texture study. Here, aeoniums, cotyledons, senecio and opuntia form a full and eye-catching hedge.

BELOW RIGHT Boulders are so important in framing and fleshing out your garden. They provide an excellent backdrop and needed blank space to showcase some of your small to midsize plants, in this case *Echinocactus grusonii* and *Aloe rooikappie*. The rock keeps the more vigorous *Aeonium* 'Zwartkop' at bay.

The gardens on these pages are all located in coastal or semi-coastal Southern California. The immediate coast can be more overcast where aeoniums thrive, but cacti bloom less. Twenty miles inland the summer heat or winter cold can cause the opposite dynamic. To borrow a phrase from astrophysics, there is a "Goldilocks Zone," say a few miles in from the coast, where there is less overcast but also fewer extremes than inland, that can be perfect for growing most succulents. Not too hot, not too cold, just right. Trial and error will allow you to fine-tune your plant selection for your little corner of the world.

ABOVE *Graptoveria* 'Fred Ives' is an aggressively clustering succulent with a muted suede-purple tone, with old leaves fading to brown. It is an excellent understory repetition plant.

ABOVE RIGHT *Senecio jacobsenii* blushes purple while a form of *Crassula radicans* blooms behind it.

TOP LEFT The usually dominant *Agave americana* is being squeezed out by an established clump of *Agave shawii*. Or maybe it is squeezing its way in.

BOTTOM LEFT A good use for deteriorating clay pots is to situate them at angles in your garden and plant them with succulents, and let their decay showcase your plants. Here a *Euphorbia tirucalli* 'Sticks on Fire' reposes behind *Aloe elgonica* and *Senecio mandraliscae*.

BELOW After your main specimens are situated, add detailed and textural plants to the garden, with repetitions of colors, such as the *Sedum rubrotinctum* in front of graptosedums.

A decomposed granite pathway winds through a lush San Diego garden by Frank Mitzel. Most of us don't have this kind of space to create a mini botanic garden, but if you do have a large lot and want to combine beauty and low-water usage, aim at something like this. You can always add plants over time.

Initially after planting your garden, you have lots of little plants and a lot of brown dirt showing. It is tempting to cover the dirt with gravel or mulch, but given time, and the right use of some ground-cover type plants, the dirt will disappear. Gravel, or river rock, do look fine in your dry river, and perhaps bordering a sidewalk or as token scatterings around the boulders. Use varying sizes of the same river rock for a natural look. You should stop adding plants to your garden when you think it is maybe 80% of what it will be. It has to grow, and you want to be able to add plants over time. Resist the urge to overplant at the start. Even though you want your garden to look finished, *do not* over-crowd, or within a year or two you will be thinning it out and wondering what you were thinking.

Succulent gardens are low-maintenance, but not no-mainte-nance. In dry seasons, they like to drink at least once a week, more in hot inland climes, and especially during the first year after installation in order to get established. The more mature the gar-den, the more you can cut back on water. Mature sentinel succu-lents can almost be treated as "vacant lot" plants here in South-ern California, such as agaves, many aloes, cotyledons, jades and many cacti. Even so, your plants will thank you and look better with summer water. My preferred method of watering is to take 15 or 20 minutes with a hose and rain nozzle after work, and do it by hand. This can be your Zen time to enjoy your plants (you can almost feel them drinking), and maybe pull a weed or two. You can also install a drip system or sprinklers if you prefer. This is usually done after planting, because you have mounds and rocks to work around.

Weeds are one constant in almost any garden. They tend to materialize in dizzying numbers after it rains. Heavy mulch may slow the weeds, but I don't like the look of the dry wood chips with the detail of a nice succulent garden. Pre-emergent weed blocker can be used, but it usually comes down to good old weed abatement at some point. Do it when the soil is wet to get the roots. Weed cloth or barrier is an option, but it is difficult to work with around rocks and mounds. It usually ends up exposed in places, the ants like the warmth these barriers provide, and I find weed seeds are air-born and will occupy the dirt on top of the barrier anyway. Can you tell I don't like working with it?

Aloe elgonica with *Portulacaria afra* 'Variegata' and *Sedum nussbaumerianum*

Agave gypsophila looking like a slithering cephalopod among bromeliads and grasses in a semi-shady succulent/tropical garden.

As you can see from the photo above, succulent gardens tend to get a bit crowded and busy. Perhaps you can just lay out your initial, tastefully limited mix of plants and be done. If so, more power to you. Done with adding plants? Then you're not a plant person.

By their very nature, succulent gardens have a lot of "look at me" plants, as opposed to tasteful, traditional gardens with a softer palette that evokes a forest glade or cottage garden. If you want to tone down the "botanic garden," every-plant-is-a-specimen look, repeat some softer groundcover succulents as well as quieter midsize silver dollar jade, cotyledons, *Agave attenuata*, and some associate non-succulents. These might include varieties of dusty millers, rosemary, artemisias, or any other drought-tolerants you like. Be sure you know their growth habits before planting. However, I'd like to defend the botanic garden approach. Take the tableau at left. That is just so much more fun to come home to.

As mentioned earlier, restraint can be difficult with the succulent enthusiast. I am a nursery owner and it is impossible not to bring home a new living creature and find a place for it in the landscape. This tends to create a crowded and ramshackle garden, but with succulent gardens, you can make it work. If you have some repetitions and theme plants, large boulders and blank spaces, you'll have the blueprint for a unique and beautiful garden—a tapestry of living curiosities that will occasionally burst into glorious bloom. If you do it right, you will have plants with touches of burgundy, red, blue, and gray in their year-round foliage, punctuated with a back-lit glow of a golden barrel or silver torch cleistocactus. If you have an aloe themed garden, you will have red and yellow flower spikes from winter into spring. Cacti will begin throwing their huge flowers in the late spring into summer. There is always something going on in the slow-motion world of succulent plants.

Although a mature succulent garden needs less water as the years pass and the plants become established, small, shallow-rooted groundcover plants may still need weekly watering to look their best. Be sure your rocks and boulders don't disappear under the canopy of plants. You may want to occasionally root some out and pull them forward for best effect.

This garden would be considered more "xerophytic" than succulent, with *Agave attenuata* and *Senecio mandraliscae* complimenting procumbent rosemary and spring-blooming proteas. Ordinarily my first comment would be "needs more succulents," but this one looks pretty nice and refined as is. When the proteas aren't blooming, the agaves provide architecture to a more subtle and shrubby, softer garden.

Can I combine succulents with other types of plants?

Yes, in fact it is encouraged. Using small grasses and other draught-tolerant plants such as artemiesias, rosemaries, sages, etc...will give your garden a more natural appearance, and soften the effect of so many succulents. They can frame your 'eye candy' specimens. Some examples of these 'associate plants' are on the following pages.

A word of caution here. Most of these drought-tolerant non-succulents will coexist nicely on the once-a-week watering schedule, but once established, can grow a lot faster and overtake or swallow up some of your smaller succulent specimens. Be aware of how large these plants will get, and plant accordingly. As with succulents, use the "big plants in back" rule, and don't be shy about trimming back on occasion. I used to have beautiful yellow-flowered Mexican marigolds (*Tagetes lemmonii*), but ended up removing the entire plant because it took over. I did the same with the beautiful blooming *Calandrinia grandiflora*, as it tends to grow faster than other succulents and also gets rangy. (Most succulents grow slowly and get showier with age.)

Can I plant a succulent garden any time of year?

This depends on where you live and what type of plants you're using, but for most Mediterranean areas such as coastal California, you can install a garden year-round. If you are in a frost-prone area, wait until the threat of freezing has passed. Along the coast, most succulents can be planted anytime, with some exceptions. Equatorial plants such as pachypodiums or alluaudias transition best in warmer weather. Transplanting can be a little traumatic for other plants, especially if they are in dormancy. Others will take off immediately, glad to be rid or their pots. Fall or winter can be a great time to plant, because winter rains give an excellent jump-start into the warm-weather growing season.

SUCCULENT GROUNDCOVERS

Senecio serpens and *Portulacaria afra minima* are both low-growing, slow spreaders that can cover ground and eventually hold back a sandy slope. If you plant a series of 6-inch plants two feet apart, they should merge within a year.

Othonna capensis 'Little Pickles' is a nice gray-green mounding groundcover with small yellow flowers. As with the senecio and portulacaria, plant 6" plants a foot or two apart and they will merge into a low carpet over time.

Sempervivum arachnoideum, if planted in repeating clumps about a foot apart, will eventually merge into a modest and interesting groundcover—provided you approve of the spider web-like nature of the plant.

Crassula 'Campfire' works as either a container or groundcover plant. In season, you get rivulets of red lava.

Oscularia deltoides, a ground-cover or cascading mesemb with pink flowers in spring.

Any groundcover-type plant you choose, in this case *Silver dichondra*, should hug the ground as close as possible and not be a plant swallower.

There are several succulents that will cover bare dirt and provide a more satisfying look than mulch or too much gravel. Ice plants tend to grow too big and overtake smaller plants. The plants on these pages are slow to cover much ground on their own, but if planted a foot or two apart, will merge into a low carpet. There are also quite a few tiny "carpet" sedums, such as *Sedum* 'Dragon's Blood' or *Sedum anglicum*, that look good for a time but in a landscape usually don't prove to be durable for the long-term. Also consider *Dymondia margaretae* (page 120) if you want a groundcover you can walk on.

This dry garden doesn't really look that dry. Designer Jeremy Spath has incorporated several varieties of grasses, sages and even small blue palms that compliment the sculptural *Aloe pseudorubroviolacea* (also inset), here looking particularly noteworthy dressed up in springtime bloom. Cardiff Library Garden, Cardiff, CA.

ASSOCIATE PLANTS

Gaillardia

Centaurea cineraria

Festuca glauca

Mariana sedifolia

Carex testacea

Euphorbia 'Ruby Glow'

Echium candicans variegated

Tagetes lemmonii

Asparagus meyeri

'Pride of Madeira', *Echium candicans*

Stachys byzantina

Calocephalus brownii, Phormium 'Burgundy'

Leucadendron 'Safari Sunset'

Leucospermum 'High Gold'

Aloe rooikappie blends in with the yellow-flowering *Calycephalus hartwegii* and other xerophytic perennials for a natural/habitat look.

Pelargonium maderense provides dramatic spring flowers but dies upon blooming.

Two of the best palms for dry gardens, *Butia capitata* in back, *Brahea armata* in front.

Santolias and lavender with *Aeonium* 'Kiwi' as hardscape borders.

Mexican sage, *Salvia leucantha*

Mexican feather grass, *Nassella tenuissima*, is dramatic and blows in the wind, but is invasive.

There are many books dedicated to drought-tolerant or Southwest gardening, with just a small portion dedicated to succulents. This is the opposite, a book of succulents giving the non-succulents short-shrift. But the palette of xerophytic companions is huge. Offered here is just the slightest sample of some of the drought-tolerant perennials that you can mix into succulent gardens. As discussed earlier, they can be more aggressive growers and usually need more frequent pruning, but if properly used can frame your garden of succulent attention-hogs with some natural softness. And even a tunnel-vision succulent enthusiast can't ignore the sublime beauty of a plant such as *Leucadendron* 'Safari Sunset', as seen above. Larger xerophytic shrubs such as leucadendrons, banksias and proteas offer subtle foliage and usually colorful and freakishly beautiful flowers in season.

Kingdom: Plantae

Phylum: Tracheophyta

Class: Magnoliopsida

Order: Saxifragales

Family: Crassulacae

Genus: Echeveria

Species: harmsii

Echeveria agavoides v. 'Ebony'

Interlude: *Echeverias & Dudleyas*

Echeverias

Echeverias are primarily Mexican in origin, with many species and even more hybrids in cultivation. They are usually soft rosettes or collections of rosettes, often in blues, blue-grays, shades of pinks and purples, sometimes hairy or striped or ruffled or occasionally covered in "carunculations," sort of a benign form of plant leprosy. You will never confuse an echeveria with a cactus. "Hens and chicks" usually refers to any type of clustering rosette succulent. In alpine or colder climates, sempervivums are usually the hens and chicks. In Mediterranean climates, this may more appropriately describe an echeveria, usually a form of *Echeveria glauca* or *E. imbricata*, which can make attractive drifts of compacted blue disks. Echeverias usually have small but showy flowers on tall stalks. Often, when one throws all its effort into multiple flowers, it will subsequently drop many of its older leaves due to the stress and shrink in size. But after the flowers, it should soon regenerate into a full plant.

Some of the most sought-after plants in the succulent world are echeverias, such as *E. agavoides* 'Ebony' or *E. cante*. The rarity of these and some other echeverias is due to their non-clustering habit. If you have a single-headed plant that is growing tall on its stalk, you can decapitate the head. Leave it suspended in an empty pot for a few weeks and you should see new root hairs forming. The stump will take just as long or longer to regenerate, but hopefully some new rosettes will eventually sprout.

Similar to echeverias are pachyphytums and graptopetalums, which differ visually from echeverias by their more fleshy, at times almost inflated-looking, leaves. They are all close allies, and can be crossed to form pachyverias, graptoverias or sedeverias. Most are prolific clumpers, easy to propagate. All these plants are native to Mexico and into Central and South America.

Pachyphytum amethystinum (in front)
Graptopetalum paraguayense (top)

Echeveria nodulosa

Echeveria 'Blue Curls'

Echeveria, crinkly hybrid

Echeveria agavoides 'Lipstick'

Echeveria imbricata (glauca?)

Echeveria agavoides 'Maria'

Echeveria carunculated, pink/green

Echeveria agavoides, red form

Echeveria 'Doris Taylor'

A hybrid echeveria, perhaps *Echeveria* 'Sinjen'

Echeveria agavoides 'Christmas'

Echeveria chihuahuaensis

An echeveria hybrid

Echeveria agavoides, variant

Echeveria 'Afterglow'

Echeveria shaviana

Echeveria cante

Echeveria peacockii

Echeveria subrigida

Dudleyas

Dudleyas are allies of the echeverias, and are one of the few ornamental succulents from California. They range from Oregon into Baja California, where many of the favored species are found. Many have a white chalky covering that provides a protection from sunburn, but from an ornamental perspective, the white really makes them stand out. Among the favorites are *Dudleya brittonii* from northern Baja, as well as its close relative from north of the border, *Dudleya pulverulenta*. There are many other desirable but somewhat more rare and obscure dudleyas, and others with very little ornamental value. Coming from a winter-wet environment, most dudleyas are best with minimal summer water. They may shrink and desiccate through the fall, and will wake up after the first winter rains. Many, such as *D. brittonii*, grow vertically on cliff faces in habitat, so if you can mimic a slope and plant them accordingly, they should thrive and present better. Flowers are usually small on long stems; not showy but sometimes interesting. Try not to touch the white leaves, as the white powder will show fingerprints for a time. Many dudleyas are still fairly rare in cultivation, particularly the single-rosette types that rarely offset. Those must be grown from seed, which is as fine as dust and difficult to collect and germinate. But as you can see from the few plants illustrated here, dudleyas can be wonderful ornamental specimens, worthy of the hunt it may take to find them.

Dudleya brittonii, aberrant form

Dudleya brittonii

A Buddha presides over a California-Asian fusion landscape featuring *Dudleya attenuata* hybrid. Dudleya flowers can be unremarkable, but these shine.

Dudleya hassei

Dudleya brittonii

Dudleya attenuata

CHAPTER 05: Theme Gardens

You may choose to design your garden around a theme. Maybe you want a garden that mimics your travels through the deserts of the Baja peninsula? You may decide to go with a color theme of soft blue or burgundy. Perhaps a subtropical jungle? This chapter presents several examples of theme gardens.

Some of us tend to collect plants that fall along the lines of particular groups, such as an aloe-themed garden. You can get fanciful and create an "undersea" coral-reef succulent garden, an alien plant invasion garden, or perhaps your theme is just "I love succulents," and you like to take home any interesting plant you find and add it to your garden. Many of us have gardens that end up following that last idea, in which case you should try to follow a couple of the principles outlined in the chapter on landscaping, in particular the idea of repetition and blank spaces.

UNDERSEA GARDENS

Some succulents bear an uncanny resemblance to undersea plants or animals. Euphorbias can look like sea stars, aloes like octopi, crested cacti draw a parallel to brain coral, and still others can mimic sponge colonies or eel grass. This is one garden where props are very appropriate, such as anchor chains or ceramic fish swimming across your terrestrial reef. Try to plant your creatures between and around as many volcanic rocks as you can to give it the feel of an actual reef.

Visitors to the nursery have always pointed out how much some succulents mimic undersea creatures. I've noticed the parallels too, so one year for the annual San Diego County Fair garden show, I decided to run with the idea. Judges like creativity. So I spent several months culling the little "sea creatures" from my inventory to use in the display. Once I started looking at plants through my "water goggles," more and more individuals presented their own versions of sponges, starfish, octopi, corals, etc. Along with some mounding

and a ton of medium-to-large lava rocks with holes and pockets to plant in, it all came together quickly. Adding some strategically placed fish-sculptures and some anchor chain, the dry land terrestrial reef came out better than I had envisioned. It garnered the major awards that year, and led me to do one for the Philadelphia Flower Show. It also gave me a moniker I would have never imagined in my youth: The Undersea Succulent Guy. Hard to explain to people.

The spiral aloe, *Aloe polyphylla* is just weird enough, it could pass as some sort of sea creature. It is from 6000 ft. elevation in Lesotho, and is difficult in warm climes. It likes cool winters, and even tolerates snow. It seems to require a difficult-to-replicate condition of bright sun, lots of water with good drainage and rich soil, and cold winters. It is one succulent that is easier to grow in the cooler climate of Northern California rather than Southern California. The spiral growth pattern, also seen in a few other succulents as well as many sea shells, is a natural manifestation of the Fibonacci sequence, always a fun topic in casual conversation. It's just fun to say *"Fibonacci."*

Undersea succulent garden at the San Diego Botanic Garden. Photo by Bob Wigand.

Photo by Bob Wigand

If you want to create your own reef, try to keep it to a smaller and visually important area. You can even create one in a bowl. As with any succulent garden, you can't have too many rocks, and elevation is key. Think of diving on a reef and descending along an undersea shelf. Use craggy lava if you can find it, and sandwich the rocks against each other, so you can plant your sea urchins in the cracks. Establish your larger plants, perhaps a large *Euphorbia resinifera*, a 'Sticks on Fire', or large crested euphorbia or cactus. Add a repetition of 'octopussy' aloes and gasterias or haworthias.

For your reef crawlers, the medusoid euphorbias are great. *Euphorbia flanaganii* is the most available, and it also looks wonderful in its crested form. Small blue *Senecio serpens* or red *Kalanchoe luciae* provide color, and any of several types of globular euphorbias or mammillarias can represent the sponge colonies. The carunculated echeverias suggest the skirts of sea slugs, and wide-bladed sansevierias can be your kelp. Take your time searching out the right plants that suit your fancy, and look for props such as rusty anchors and chains, old lobster traps, and if you want to be truly authentic, maybe a few rusty beer cans.

Crested plants are also integral to an undersea garden but are hard to come by. Before you plant your coral reef, take some time to stockpile plants. They'll be happy in the pots you bought them in until you are ready to plant your garden.

Stapelia variegata

Euphorbia flanniganii

Crassula 'Hobbit'

Dyckia 'Brittle Star'

My vote for the "octopus" agave, which usually refers to *A. vilmorineana* would be for the twistier *Agave gypsophila*.

A stem crest of *Euphorbia flannigani* appears ready to snatch up any creature that swims close enough.

A particularly blue form of *Myrtillocactus geometrizans* undulates up from its rocky lair.

Photo by Bob Wigand

143

Denizens of the Dry

Euphorbia obesa crest

Euphorbia obesa

Stapelia grandiflora

Euphorbia polygona

Euphorbia esculenta

Aloe vanbalenii

Euphorbia alternans

Stapelia gigantea

Carunculated echeveria

Euphorbia woodii, similar to *E. flanaganii*

Anchor at rest among *Euphorbia obesa*

Dyckia marnier-lapostollei

Unidentified mammillaria which has contorted itself into a brain cactus, or "brain coral".

Photos this page by Bob Wigand, taken at the San Diego Botanic Garden. Design by Bill Teague and Jeff Moore.

DESERT GARDENS

Desert gardens can have the look of a true natural habitat, albeit one that is spartan and dry. This aesthetic may appeal more to an amateur naturalist or desert traveler than a homeowner looking for a colorful and softer yard. These plants do lend themselves well to gardening in the desert Southwest, where many of the softer succulents have a tough time with the extremes of hot and cold. The Mojave desert of the southwest U.S. also offers a wonderful array of succulents and cacti, such as the ocotillos, ferocacti, opuntias, etc. If you've experienced the deserts of the U.S. and northern Mexico, this is a nice way to represent a piece of them in your yard. They are also a great display area for the rock hounds among you. If you are lucky enough to have petrified wood, this is the place to show it off.

The images seen on the two pages here are of the Mojave Rock Ranch (http://mojaverockranch.net/), an extraordinary 'rock and bottle' house in the Joshua Tree region of the Mojave Desert in California. Owners and designers, Gino Dreese and Troy Williams, work with the cactus and succulent options available to them in the high desert. The incorporation of glass bottles into walls made of native rock, combined with rusty tools, tile, and driftwood is a wonder of architecture in itself, but for our purposes, the plants are the thing. Out of necessity, this garden lacks the softer succulents we use closer to the coast. But they would likely look inappropriate in this context anyway. These are the sharp and sculptural backlit beauties of the desert. Golden barrels and chollas, even some agaves can have a glowing halo of spines in the low winter sun. Ideally suited for desert extremes, many of these plants will live along the coast as well. Other American Southwest desert gardens to visit include The Living Desert in Palm Desert, CA, the Desert Botanical Garden in Phoenix, AZ, and the Arizona-Sonora Desert Museum in Tucson, AZ.

Desert Gardens photos taken by Gino Dreese and Maureen Gilmer.

SUBTROPICAL GARDENS

Some people can't shake the desert-cactus-sparse-look, when they think of succulents. In Southern California, we've figured out how to combine succulents with xerophytic tropical (pardon the oxymoron), or at least tropical-looking plants. As you can see here, there are many colorful bromeliads that can combine with lush succulents to go with your tiki decor. There's no reason you can't combine aloes and aeoniums with palms, staghorn ferns and all manner of bromeliads to create your subtropical getaway.

INVASION OF THE XERIPHYTES

Many succulents have a decidedly alien appeal. They look like they may be from outer space. These ceramic aliens arrived at a garden show to represent their tribes from far-flung galaxies. The Echeverians, Deuterochonians, Euphorbians and Cactaceans have found our dry climate an ideal replacement for their dying worlds. Fun with plants! These ceramic alien pots were made by Jim Moore.

FIRESAFE GARDENS

Succulents are about as fire-safe as any type of plant there is. By their very nature they are sort of big water balloons surrounded by thick leathery skin. However, dead and dried portions of succulents can catch fire. Some people have surrounded their Southern California homes with ice plant, which doesn't exactly burn, but as recent wild fires have shown, their under-canopy of dead leaves can smolder for days. One of my customers built a living wall between his house and a canyon with large opuntias and green pencil euphorbias. The devastating San Diego Cedar fire in 2004 caused him to evacuate. When he returned, he found that a low wall of fire burned up the hill until it hit his succulent fence, and in his words, "wanted no part of it." The wall was blackened by the burn, but was still green on the inside. He partially credits his succulents for saving his house. The sprinkler on the roof might have helped too. Driving past fire-charred areas, within weeks some of the first signs of life pushing out of the charcoal were new shoots of what looked for all purposes to be incinerated yuccas and agaves. Succulents sometimes don't know when to give up.

GENUS-THEMED GARDENS

Some of us have a fascination with a particular plant group or genus. In my visits with customers and suppliers, I've seen cactus-themed gardens (very popular in desert climes), agave-heavy gardens, aloe gardens, and others that are not so much genus-oriented as they have a "feel", such as the soft and pastel-type plants, including echeverias, sedums, crassulas and aeoniums (no spikes!). One scan of my yard will tell you I'm an aloe guy. As mentioned earlier in this book, as long as you use some repetition, particularly with some small to medium varieties, as well as repeated ground-cover types, you can have what appears to be a kind of fantastically varied habitat garden. Peak flower time for an aloe garden is December through February, although with a good collection, something will be blooming year-round.

COLOR-THEMED GARDENS

With succulents, you can create a red garden with red aloes, a yellow-variegated garden, or a blue garden with agaves and senecios as seen here. Blues and whites with splashes of burgundy make for great garden color.

Kingdom: Plantae

Phylum: Magnoliophyta

Class: Magnoliopsida

Order: Rosales

Family: Crassulacae

Genus: **Crassula**

Species: ovata

Crassula susannae hybrids

Crassulas, Senecios & Sedums

Crassulas, senecios and sedums represent very diverse groups of plants, many of them drifting out of the category of succulents and into more cold-hardy plants. Presented here is just a token offering of each group. Most are non-spiny, more "gardeny" in appeal or foliaceous in appearance. They are generally softer and occasionally colorful or artistic plants. There are also quite a few cold-weather sedums that don't quite cross over into the succulent category, usually a little more thirsty and less sun-tolerant.

Crassula 'Baby Necklace' (Crassula rupestris v. marnieriana), beginning to envelop an echeveria.

Crassula falcata

Crassula capitella ssp. thyrsiflora

Crassula nudicaulis

Crassula radicans, large form

Crassula coccinea v. 'Campfire'

Crassula pubescens

Crassula capitella 'Red Pagoda'

Crassula ovata 'Hummel's Sunset'

Crassula deceptor

Crassula perforata

Crassula ovata 'Wavy Jade'

Senecio amaniensis

Sedum morganianum

Sedums, senecios and crassulas are usually excellent plants for container culture, and some are well-suited for hanging pots or cascading situations. Some of these are *Sedum rupestre*, *Sedum nussbaumerianum, Sedum morganianum, Crassula* 'Baby Necklace', *Senecio jacobsenii* and *Sedum morganianum,* a.k.a. *Sedum* 'Burrito'.

Senecio jacobsenii

Sedum dasyphyllum

Senecio haworthii

Senecio vitalis (back), *S. mandraliscae* (front)

Sedum rubrotinctum 'Aurora'

Sedum rupestre 'Angelina' with *Sedum dasyphyllum*

Sedum rubrotinctum 'Pork and Beans'

Crassula rupestre, green and blue forms

Senecio serpens

Sedum nussbaumerianum

Crassula 'Hobbit', one of the varying forms of the ubiquitous 'jades' that makes an excellent bonsai or dwarf landscape tree.

CHAPTER 06: Succulent Bonsai

In the introduction to this book, I alluded to several 'light bulb' moments, when the spell of succulents really took hold somewhere in the back of my brain. The final step off the cliff for me was seeing Rudy Lime's award-winning collection of bonsai succulents at the Del Mar Fair in 1984.

Traditional bonsai is one of our highest forms of ornamental horticulture, but Rudy Lime's display of one man's manipulation of bizarre and (to me) unknown botanical curiosities into a succulent medium of living art was the coolest thing I'd ever seen. I found a new calling that day.

You can't help but notice the bonsai potential in many succulent plants. Harsh elements in nature can cure many habitat plants into wind-sculpted miniature mimics of real or imagined larger plants. Because most succulents are shallow-rooted and can be coaxed into small pots, and are low-effort and low-water plants, this is sort of like traditional bonsai but much easier, with the added appeal of an alien-like or cartoonish nature. When Rudy Lime arrived in San Diego from the Philippines, he immediately translated his love of bonsai into this new succulent medium. He entered some of his succulent bonsai into a traditional bonsai show, and walked away with most of the ribbons. He found himself relegated to a new division the next year. The traditional spruce/pine/maple bonsai guys didn't get it. Call it what you will, this miniature presentation of remarkable plants is one of the most creative and sublime expressions of ornamental horticulture. Make an effort to check out the bonsai succulent display at a cactus-and-succulent show.

Rhombophyllum dobaliforme

Portulacaria afra, variegated

Large succulents such as saguaros or dragon trees are obviously long-lived, as it can take at least a hundred years to reach such magnificent size. You don't generally think of the smaller plants as being capable of living to such an advanced age, but they can. Take the 'African bonsai' (*Trichodiadema bulbosum*) at right. This is one of a group of plants I acquired from a lady in 2013 who had been nursing them along since 1961! I was amazed they were that old, but it turns out that in fact they were older.

Following is a paraphrase of her letter to me about them:
"I've had the "Ito" succulents since 1961 (the Itos had a nursery in Leucadia, CA in the '50s and '60s). Mrs. Ito bought them in 1941, prior to Pearl Harbor. When she and her family were interned in North Dakota, a neighbor watered her plants until her return. When I first saw these, they were all—dozens of them—lying on their sides, still in 2-inch liners, with roots out the liner holes, firmly rooted in the ground against one of the Ito glasshouses. I bought a dozen at 50 cents each, and gave a few away over the years."

'African bonsai' (*Trichodiadema bulbosum*)

There are a number of plants referred to as 'jade', which usually are varieties of *Crassula ovata* or *C. argentea*. Some of the popular cultivars are *Crassula* 'Hobbit' (above) and *Crassula* 'Gollum', which both lend themselves to bonsai treatment without the usual fuss that "real" bonsai require. All the cut pieces can be re-rooted; there is seldom any reason to try to grow jade from seed. Jades are among the easiest of all succulents to grow. I joke that you could put one through a wood-chipper and whatever comes out the other side would grow.

ABOVE is a *Crassula* 'Hobbit' (although some will argue that), which had been stunted in a small plastic pot, then had most of its stems hacked back. The resulting new growth gives it a fat baobab tree look with a tight canopy.

ABOVE RIGHT An old *Tylecodon paniculata*, pot-bound for over twenty years (it has been repotted a few times, into progressively slightly bigger pots). Had it grown this amount of time in the ground, it would likely be a 3- or 4- foot high bush, as opposed to the fat 10-inch tall little Buddha it has become. Cotyledons and tylecodons are winter growers, opposite season to many superficially similar succulents. It leafs out in October; the leaves yellow and fall in the spring, usually leaving long flower stalks through the summer months.

BELOW RIGHT *Euphorbia misera* is generally a tangled shrub in its native habitat from coastal San Diego County in California into northern Baja. Old plants develop a significant trunk structure, and are occasionally groomed by the elements into a natural bonsai shape. In cultivation, an artistic plantsman such as Peter Walkowiak can trim and stage a small succulent bush into a miniaturized piece of art.

ABOVE LEFT This old fat-plant example of *Pachypodium lealii-saundersii* is a caudiciform that takes on the characteristics of an old bonsai as it ages, with little training required. It is winter-deciduous, flowering from late summer into early fall. Shelter from winter rains when possible.

ABOVE MIDDLE The *Mestoklema tuberosum*, a mesemb from the "expose your roots" school of bonsai. This mangrove-style creature can look like it is ready to walk right out of its pot. It has delicate orange flowers.

ABOVE RIGHT The Madagascan tree succulent, *Uncarina grandidieri*, forms a small blooming deciduous tree in the landscape, and also lends itself to bonsai training. It flowers in the warm months and offers an unusual, fuzzy yet spiny seed pod after flowering.

BELOW LEFT is a plant most people would be surprised to learn is in the cactus family. It is *Pereskia aculeata*, a leafy and woody primitive cactus relative (it does have spines) that can be trained into a bonsai shape.

BELOW RIGHT An example of the elephant bush, *Portulacaria afra*, trimmed back hard in the early stages of bonsai. This type of plant can be hacked back to barest of stumps, and it should regenerate new foliage that can be trimmed or wired into proper bonsai shape.

Adenium obesum

Ficus palmeri 'Mexican rock fig'

Pachycormus discolor

Fockea edulis in all its twisted-root/trunk glory. Keep a starter plant in a plastic pot for a long time and you'll find a twisted caudex ready to be raised and exposed. If it doesn't exactly look like a bonsai, it at least looks like a vegetative alien creature.

TOP RIGHT and ABOVE are two examples of the elephant trees of Baja, *Pachycormus discolor*. It takes a long time to develop the characteristic fat body.

Basic bonsai principles apply with succulents. The main differences are in their water, soil and light requirements. These are still succulents, after all, just with a different presentation. As with traditional bonsai, you want to make your plant look like a miniature version of some imagined larger tree. Caudiciform plants in particular can make excellent bonsai, owing to their large bodies, which can resemble the fat trunk of perhaps an African baobab tree. Judicious pruning, wiring, and staging can help you create your miniature tree. As most succulents are shallow rooted, a shallow bonsai or hand-built pot is appropriate.

Plants on this page courtesy of Julian Duval.

Rudy Lime has been shaping succulents into miniature trees for years. Seen here are some examples of how he applies traditional bonsai techniques to some non-traditional material. Above are specimens of *Portulacaria afra*, a very ordinary jade-like succulent that normally grows shrubby. Trim the foliage and smaller branches way back until you have the basic shape of the tree you envision. You can use bonsai wires, as seen in the bondage shot at left, to help pull it into the shape you want. Over time, new growth gives it a more lush appearance.

Plants on this page courtesy of Rudy Lime.

This is the fairly rare *Fouquieria purpusii*, a greener and smaller version of the boojum of Baja, *Fouquieria columnaris*. This plant has a natural dwarf appearance, becoming a caudiciform with age. Very little pruning is needed; the hardest part is just locating one of these plants. Find the proper pot, add rocks, and let it age.

An example of the Madagascan tree succulent, *Operculicarya decaryi*. Owing to its stout trunk, tiny leaves and overall pliability, this may be one of the best of all candidates for bonsai. New leaves are a dark copper color, becoming green over the summer, then fall off with autumn and winter dormancy.

Kingdom: Plantae

Phylum: Streptophytina

Class: Magnoliophyta

Order: Saxifragales

Family: Crassulaceae

Genus: **Kalanchoe**

Species: luciae

Kalanchoe luciae

Kalanchoes

Kalanchoes are characteristically "soft" plants, often coated in a smooth fuzz, or occasionally forming hundreds of little plantlets on the leaves. You won't find anything remotely resembling a spine on a kalanchoe. The most ubiquitous is *Kalanchoe blossfeldiana*, sometimes called the "supermarket kalanchoe," as it is common in the house plant trade. It is a compact, green, fleshy-leaved plant with showy flower clusters. However, they usually never bloom again quite as well as that greenhouse-forced plant you first bought.

Kalanchoes (usually pronounced "kal-en-koey," or "-koa," not "kal-AN-cho") are mostly Madagascan plants that don't like cold weather. They are usually the first droopy casualties of even a light frost. However, in warm weather they can become attractive specimens, particular the "Napoleon's Hat," *Kalanchoe beharensis*. (If you see a picture of Napoleon with his hat, you might agree that a better name based on its leaf shape would be the Paul Revere hat). There are several "mother of millions" varieties that drop little plantlets that will root below the parent plant. This can make it a bit of a weed, but it is an attractive weed that you can give to your friends.

Kalanchoe luciae 'Fantastic'

Probably the most popular kalanchoe in the succulent trade is the "flapjack," *Kalanchoe luciae* (formerly *K. thyrsiflora*, which is actually a related, less red relative). The red-orange form on page 167 is the most common, but new variegated forms are appearing on the market.

BOTTOM LEFT is the variegated form known as 'Fantastic' for obvious reasons, and the super-variegated, almost pure yellow form of the same plant is shown above it. These versions are fairly new to the trade, mutations that popped into existence and wisely propagated by an attentive grower. Whether through the discovery of a new species, an intentional hybrid, or a spontaneous mutation, there is always something new in the world of plants.

Kalanchoe beharensis

Kalanchoe orgyalis

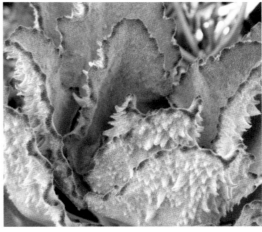

Kalanchoe beharensis v. 'Fang', a cultivar I sometimes call the "Feed me Seymour" plant, a.k.a. the Audrey 2 from Little Shop of Horrors.

Kalanchoe thyrsiflora

Kalanchoe tubiflora 'Pink Butterflies'

Kalanchoe blossfeldiana 'Supermarket Kalanchoe'

Kalanchoe daigremontiana

A garden cultivar, likely a hybrid of *Kalanchoe marnieriana*

Kalanchoe gastonis-bonnieri

The felt-leaved 'Napoleon's Hat' *Kalanchoe beharensis*

Kalanchoe marmorata (top) *and Kalanchoe tomentosa* (bottom)

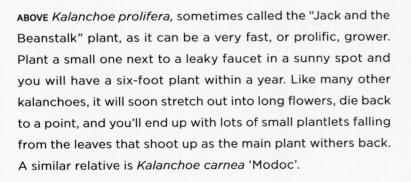

ABOVE *Kalanchoe prolifera,* sometimes called the "Jack and the Beanstalk" plant, as it can be a very fast, or prolific, grower. Plant a small one next to a leaky faucet in a sunny spot and you will have a six-foot plant within a year. Like many other kalanchoes, it will soon stretch out into long flowers, die back to a point, and you'll end up with lots of small plantlets falling from the leaves that shoot up as the main plant withers back. A similar relative is *Kalanchoe carnea* 'Modoc'.

Kalanchoe fedtschenkoi variegata

CHAPTER 07: Crests

With such a variety of forms, there are many niche interests that can draw you in. You can be attracted to a particular genus, such as the aloe group, or perhaps you like variegated plants, or certain flowering cacti. The phenomenon of cresting is a naturally-occuring mutation that creates a common visual kinship in plant families as diverse as cacti, euphorbias, or aeoniums. It can turn an ordinary plant into an extraordinary specimen.

Cresting, or fasciation, is still a bit of a mystery. The growing meristem of a plant will occasionally grow in what is called a fasciated, or crested form. It may be a type of benign virus, or something analogous to that. Whatever the cause, no one has figured out yet how to make a plant crest, but once one does, it can often be cut up into more crests. Vegetative propagation is the only way to create new crests.

This is a phenomenon that is fairly common in succulents and cacti, but rarely seen in other plants. It will occasionally show in certain flower stalks of non-succulents, and rarely in plants such as Echium and the 'pygmy date palm,' *Phoenix roebelenii*. Most people consider it a wondrous mutation, but some see it as a grotesque abnormality, which is hard to argue with. If you appreciate the bizarre and exotic, these plants may be for you. Many collectors specialize in these mutants, along with rare variegates. Of course, crested variegates are even better!

Crests do occur rarely in nature. There are some spectacular crested saguaros in habitat in Arizona.

The term "brain cactus" can be applied to many crested or crenulated plants, but the most "brainy" is probably *Mammillaria elongata* in its crested form. However, I've found it is difficult to grow outside of a greenhouse. Every one I've ever tried to grow outside has eventually melted away.

Spiral crest of *Pedilanthus macrocarpus*

FACING PAGE If you can make a case that a living thing can be a sculpture, I submit the specimen at right on page 173. In my 20-plus years of running a succulent nursery, I can't say I've acquired a more magnificent specimen than this crested *Myrtillocactus geometrizans*. When a plant's programming goes haywire and causes it to convolute, there is no way to anticipate the form it will take at its glacial pace. Occasionally there will be a reversion to normal growth, which probably should be removed to encourage growth back into the crest. This plant shows no reversions, just a continual building upon itself like lava oozing up to the surface. This is a Picasso of a plant, or maybe a Dali. I'm glad I got a picture before it went to a new home. I even broke my usual practice and put a "don't buy me" price on it, and it sold anyway. I try not to get attached to plants because I make my livelihood selling them, but this is the one that got away. I hope it's still doing well. I can't talk about this one anymore, too many memories, I have to turn the page...

TOP PHOTO A 'stem crest' of *Euphorbia flanaganii*. **LOWER MIDDLE** is considered to be a 'leaf crest' of the same plant, and **LOWER LEFT and RIGHT** are examples of
E. flanaganii in the regular form.

Euphorbia suzannae crest

Echeveria 'Big Red' crest

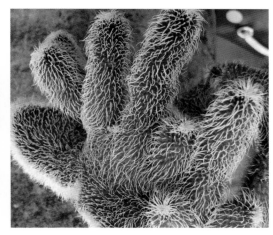

Hoodia crest

Lophocereus williamsii crest

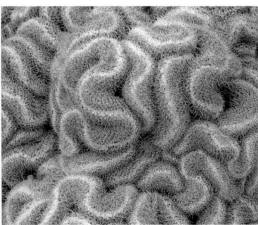

Mammillaria geminispina crest

Euphorbia 'Zig-Zag' crest

Mammillaria spinosissima crest

Hildewintera aurespina crest

Euphorbia trigona crest

Carnegia gigantea. A monstrose form of the saguaro cactus that looks like it was created drip-castle style.

Lophocereus schottii, the totem pole cactus. This is the monstrose form, which is similar to fasciation, where plants typically grow bumpy and convoluted.

A *Cleistocactus strausii* signals "A-OK"

Cephalocereus senilis crest

Euphorbia lactea 'White Ghost'. An albino crest with reversions back to normal growth.

Euphorbia mauritanica crest

Euphorbia kibwezensis, crested form

Kingdom: Tracheophyta

Phylum: Magnoliopsida

Class: Malpighiales

Order: Euphorbiales

Family: Euphorbiaceae

Genus: **Euphorbia**

Species: ingens

This "tree" is *Euphorbia ingens*. It grows relatively fast in Southern California and is our alternative to the saguaros of the Southwest desert, serving as one of the large sentinel plants in many gardens. The flowers are small and the plant is showier after the flowers grow into seed pods as seen here. On hot days they can audibly pop and spread seeds nearby.

Euphorbias

"Euphorbia" is a good default guess when trying to name a succulent. There are just so many types in all shapes and sizes. Many are columnar cactus-like in structure, others mimic clustering mammillaria cacti, some are mounding pillow-shaped affairs, and there is a huge group of leafy types, some of which exhibit little or no succulence and can grow in cold climates. There are even some tiny euphorbia weeds. What they superficially have in common is a caustic white latex sap, and similarities in their small flowers. Most succulent euphorbias are from Africa and the Old World, but there are New World euphorbias and close relatives.

Euphorbias often have a waxy feel to their skin, and if they have spines, they usually aren't very dangerous. What can be dangerous is the sap, although the danger is greatly exaggerated. Most people don't have a skin reaction to the sap, although some are allergic. What you do want to avoid is sap-to-eye contact. It stings, or worse. I have had euphorbia sap drip into and dry in my hair, only to leak into my eyes in the shower that night. Always have eye drops handy. Regarding children and pets, the latex is poisonous, but tastes so bad that the tales of internal poisoning I've heard are few and non-fatal. The ingested amount it would take to be harmful would be awfully hard to swallow. The Christmas poinsettia has been battling this stigma for a long time. My kids grew up around euphorbias and some nasty cacti and have learned to have a healthy respect for them and keep their distance. But the safe answer is don't get a euphorbia if you have doubts. I've learned not to try and convince a doting grandma not to worry so much.

Some euphorbias, particularly the columnar and 'cactiform' plants, will occasionally exhibit spotting or streaks of brownish scars, particularly at the joints, occasionally running up some arms. The reasons for the spotting can be anything from a fungus or mildew to extremes of heat or sun from an occasional weather event alien to its' native environment. There's nothing you can do to make the marks go away, but if the plant still is growing new tips, new growth may make the scars appear minor over time. Regarding the wood-like skin at the joints, this may be an example of 'lignifying', or becoming wood-like, for structural integrity. They do that.

Euphorbia clandestina, a truly alien looking little creature that can resemble a succulent palm from an imaginary galaxy. Or maybe those little green "flowers" are really alien eyes.

Euphorbia fruticosa in flower

Euphorbia stellaspina

A cluster of *Euphorbia polygona* 'Snowflake'

Euphorbia bougheyi, showing the reptilian markings reminiscent of snakeskin.

181

ABOVE the Christmas poinsettia (*Euphorbia pulcherimma*) in its present hybridized form is almost impossible to grow anymore as an in-ground plant. The landscape poinsettia seen here is a cutting from an old original hybrid known as 'Magdalena Ecke', closer to the original species stock. If you live in Southern California, try to find some old stock and establish a cutting. Old material is rarely available in the nursery trade. The red "flowers" are actually leaves; the actual flowers are tiny and nondescript.

TOP LEFT *Euphorbia lambii* from the Canary Islands is at home in Southern California.

BOTTOM LEFT one of the most familiar "cactiform" euphorbias is *Euphorbia trigona*. It grows fat and full in the sun and has leaves in the growing season. It is adaptable to bright, indirect light indoors or on patios, and is therefore almost acceptable as a "house plant."

FACING PAGE *Euphorbia grandicornis* displays small yellow flowers typical of the genus.

An unidentified medusoid euphorbia in bloom.

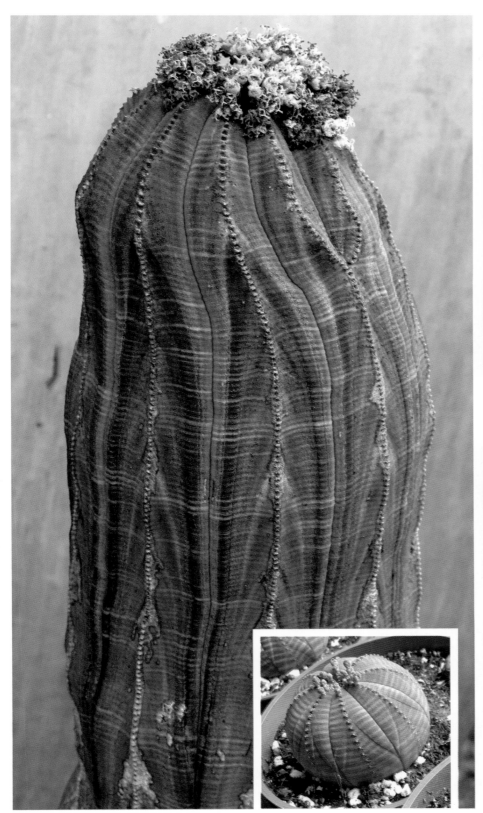

Euphorbia ammak, variegated (crest)

Euphorbia ammak, variegated

Euphorbia obesa, a very old and vertical specimen. (INSET) The more typical younger, baseball-shaped version of the plant.

Euphorbia milii, a compact form of crown of thorns known as *E.* 'Dwarf Apache'. These plants seem to bloom almost year-round.

Unidentified cactus-shaped euphorbia

AT LEFT A large tree-cactus type of euphorbia, likely the less-seen green form of *Euphorbia ammak.* This plant was a 3-footer with a few small arms when planted 10 years prior to this photo. Now it is over 20 feet tall and nearly as wide. These types can grow big relatively fast, so take that into consideration when planting. They also seem to defy gravity and high winds, although arms can occasionally break and fall (and sometimes root on their own where they land).

Euphorbia balsamifera

Euphorbia pugniformis

'Thai hybrid' giant form of Euphorbia milii

Euphorbia bupleurifolia suggests a miniature cycad, or an alien plant ambassador.

Euphorbia anoplia

A close relative to Euphorbia obesa is Euphorbia obesa ssp. symmetrica, a small and sculptural little ball of a plant.

AT LEFT Euphorbia tirucalli 'Sticks on Fire' is an aberrant form of the ubiquitous "pencil tree," the green form of E. tirucalli. It was introduced into cultivation in Southern California in the mid 90s and has become a staple "wow" plant in many gardens. The color is highlighted in the new growth at the tips, and seems to be at its fiery peak in the cool winter months. In most smaller landscapes, you really only need one as they can overpower the rest of the garden. Note: this euphorbia has a tendency to fade at times, particularly in warm weather, into a washed-out yellow-green; however, the color usually comes back with new growth during the winter.

CHAPTER 08: Variegation

Whether you were aware of the term or not, you have seen variegation. Simply put, it is a form of striping, usually yellow or white stripes (or occasionally whole leaves or plants), rather than the usual green. It occurs throughout the ornamental plant world, seen in many forms of shrubs, tropicals, bamboos, grasses, etc.

With succulents, variegation really can make a plant "pop." As with cresting, variegation occurs rarely in nature; but in cultivation, variegates are culled from the herd and multiplied for our pleasure. It can turn an ordinary-looking plant into a show-stopper, as you can tell from the photos. Certain rare variegates (particularly the miniature agaves at present) command good money. Tissue culture is now bringing more to the market. There is a school of thought that variegation makes a plant weaker in full sun due to a lack of chlorophyll, but I've seen quite a few variegates do fine in blazing inland sun, such as the variegated forms of *Agave americana* or *Euphorbia ammak*, and quite a few of the variegated aloes and aeoniums. However, if a plant produces an entirely yellow pup, it usually is difficult to grow, although it seems to do fine if still attached by its umbilical to the parent plant. Variegation is primarily yellow, occasionally white or pinkish. A while back, I had an aloe collector from South Africa, the home of most aloes, visit me at my nursery. He was amazed to see we had *Aloe arborescens* and *Aloe saponaria* in the variegated form in California. He hadn't seen them at home. So he took some long-lost little mutants back to their homeland to be re-introduced as returning royalty. I was happy to get a shipment a few months later of some unusual flower-colored aloes from Africa. We plant geeks have fun that way.

A variegated *ferocactus*

Agave decipiens variegated

Aloe saponaria variegated

Stare at the plant, keep staring...cross your eyes...keep staring...you are getting sleepy...

Agave lophantha medio-picta (?)

TOP LEFT Highly sought by collectors, this is the variegated form of *Agave parryi v. truncata*. This plant is still rare enough that a small 8" specimen commands several hundred dollars.

ABOVE A cluster of *Agave weberi* in the variegated form. At first blush this plant resembles the more common *Agave americana variegata*, but this one has a brighter variegation, less spination, and a subtle glaucous sheen to the skin. It also maxes out at around 4 feet, as opposed to *A. americana*, which can easily go 6 feet in width and height.

MIDDLE LEFT *Agave americana* 'Quito Form'

FAR-LEFT A striated/variegated form of *Agave desmettiana*. Many plants, including agaves, have more than one variegated form. They may have a mid-leaf yellow or white stripe; green middle with yellow on the margins; tricolor variegation, usually yellow and two shades of green; both a bright yellow and pale yellow form; or they may have striated streaking as in this specimen. That just makes life more fun for collectors. They love finding a new oddity and posting a photo online to turn their peers green with envy (or, *variegated with envy*).

BOTTOM NEAR-LEFT The aloe at left is a true oddity. It is an unusual tree aloe hybrid, sometimes called 'Goliath', that is likely a cross of *Aloe barberae* and *A. vaombe*. A variegated specimen is unheard of, let alone a large almost purely variegated specimen growing in-ground. This plant looks like it would be at home in a tropical garden, but it was discovered and isolated at Rancho Sole-dad Nursery near San Diego, California.

ABOVE *Agave macroacantha* variegated

ABOVE LEFT A couple of variegated agave creatures appear to be crawling out of the jungle, showing contrasting patterns of variegation. The larger one in back appears to be a more common form of *Agave americana* with the typical green mid-stripe. The unidentified agave in front exhibits the "medio-picta," or mid-stripe form of yellow inside. Both of these large agaves exhibit some of the relaxed, octopus-like leaves as opposed to the more frequently seen upright, more rigid habit.

BELOW LEFT *Agave angustifolia* in a high-yellow state of variegation. Pure variegates can be weak growers, but this one may have just enough green in the mid-stripe to keep photosynthesizing.

BELOW *Agave macroacantha* with variegated edge-stripes. This is a variable agave and is the parent of some nice hybrids, particularly *Agave* 'Royal Spine'. The form shown here is a bluer and more truncated version than the *Agave macroacantha*, shown in a striated phase in the photo above.

A striated form of *Agave celsii*, or perhaps a celsii hybrid. This agave at first glance looks like *A. attenuata*, but has subtle teeth along the leaf edges and a stiffer leaf than *A. attenuata*.

Aloe nobilis variegated

Agave attenuata variegated

Astrophytum myriostigma variegated

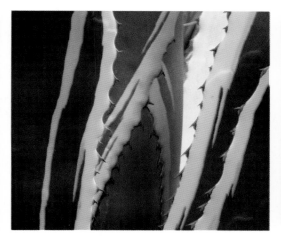

Agave ferox variegated

Kalanchoe luciae 'Fantastic'

Aeonium 'Sunburst', crested form

Crassula ovata variegated

Haworthia limifolia variegated

Agave applanata in a muted variegated form

We are currently in a heyday of new variegated plants being introduced, none more so than the agaves.

ABOVE *Agave* 'Joe Hoak' is a faded yellow-white melanistic form of *Agave desmettiana*, distinct from the common green/yellow form or the much more rare striated variegation on the previous page.

ABOVE RIGHT is the *Agave victoriae-reginae* 'Albomarginata'. It is very highly treasured among enthusiasts, and commands a high price. Tissue culture may eventually make it more available. It seems to hold its color in the elements, unlike the yellow/lime green form seen on the next page, which tends to fade in the sun but looks stunning in a greenhouse.

BELOW RIGHT is a highly variegated example of common *Aloe vera*. Unfortunately, attempts to grow this plant out of a greenhouse environment have caused it to burn and fade.

Agave victoriae-reginae variegated

Agave americana medio-picta alba

Agave 'Blue Flame' variegated

Agave gypsophila 'Ivory Curls'

Kingdom: Plantae

Phylum: Tracheophyta

Class: Magnoliopsida

Order: Caryophyllales

Family: Aizoaceae

Genus: Mesembryanthemum

Species: crystalinum

Delosperma sphalmanthoides

Mesembs

"Mesemb" is short for mesembryanthemum, juicy little African plants that occasionally mimic rocks. Most have bright fall to winter flowers in yellows, oranges, pinks, whites, etc. The most common mesembs around the world are the "ice plants," which are carpet-forming ground covers that annually explode into a solid mass of retina-burning color. Mesembs are succulent in every sense of the word. They often live between and blended in with their rocky surroundings, expanding and splitting when they get rain, then desiccating and shriveling in the dry months. There are only a relative handful in cultivation outside of their mostly South African homeland. Hopefully some of their relatives will make it into the nursery trade.

Lithops aucampiae

A cluster of various lithops

Everyone that has tried to grow a lithops has killed a lithops ("lithops" with an "s" is the singular form of the word). They are sensitive to too much water during dormancy. The best advice is to keep them drier than your other plants. They want to be dry through the cold winter months, so don't grow them in the ground if you have winter rains. They also want to be dry during the hottest part of the summer, so water sparingly, mainly in fall and spring. When you see one shriveling, resist the urge to water; the new "leaves" are in the process of consuming the old leaves.

Conophytum in bloom

Titanopsis calcarea

Form of cheiridopsis, a sort of blue "Gumby" plant from South Africa.

Glottiphyllum longum

Argyroderma

Various "living rocks"

'Split rock,' *Pleiospilos nelii*. This resembles a giant lithops, but is much easier to grow.

Cheridopsis

Carpobrotus quadrifidus

There are many forms of spring-blooming "ice plant", such as the pink carpet or orange forms of lampranthus shown here. They are aggressive growers, so don't plant in the company of smaller succulents. Lower right is the true "ice plant", *Mesembryanthemum crystalinum*.

Are some of the non-native succulents invasive?

Only a couple of them are a concern in Southern California, nothing like the kudzu problems in the South. A few types of ice plant, mainly *Carpobrotus* "freeway iceplant" can spread via bird or possibly other methods of seed distribution to become pests in the native chaparral. Some coastal lagoons also have the unusual *Mesembryanthemum crystalinum* creeping around, although it doesn't seem as invasive. Most succulents are very good about staying put. Those that do "pup" or even set seed without outside help tend to only occupy their immediate area, and even that takes many years. You will occasionally see an agave colony making a 30-year creep from a backyard down an adjacent canyon, but even that is not at much expense to native flora. If you do see non-native succulents inhabiting a stretch of native flora, they should be removed. If it is in a narrow strip already horribly corrupted and surrounded by development, there may be no point.

'baby toes' *Fenestraria aurantiaca*. Little windows in the tips allow light into the interior of the leaves for photosynthesis.

CHAPTER 09: Caudiciforms

Obesity generally isn't such a good thing for humans. You can make the argument that storing up extra fat can help in lean times, but I don't think many doctors would buy that excuse. However, with some succulents morbid obesity is a good thing. They really can store up water in their bodies for dry times, and an added benefit, at least for us, is the visual appeal that results, from cool to comical.

"Caudiciform" is a descriptive term for fat plants with a visual similarity, but is not in itself any particular genus or plant family, such as aloe or agave. Caudiciforms, also called pachyforms (or in the non-Latin sense, "fat plants") have an appeal all their own. These slow-growing succulents generally store water in their bodies, referred to as the caudex, or occasionally in subterranean roots, which can be raised and displayed partially above-ground. Their swollen appearance often gives them the feel of miniature baobab trees, hence many are popular in succulent bonsai. Others, such as the beaucarneas and dioscoreas, or succulent trees such as *Brachychiton rupestris*, don't convey bonsai as much as an alien presence, or a real-life representation of something from the world of Dr. Seuss.

Most caudiciforms must be grown from seed and take years to develop their character, and therefore are usually expensive and/or hard to come by. They are among the first plants noticed in a succulent collection, the Jabba the Hutts of their little empire of beauties and weirdlings.

A mature *Beaucarnea stricta* as seen in its entirety.

A mature *Beaucarnea stricta* with caudex textured like a tortoise's shell.

Dorstenia gigas

Cyphostemma juttae

Cyphostemma juttae

Brighamia insignis

Pachycormus discolor

Fockea edulis

Bombax ellipticum

Caudiciforms are succulence to the extreme. The fat *Pachycormus discolor* on the opposite page is a denizen of the dry Baja desert; its body is a living canteen which allows it to leaf out long after the last drink of rain. It can become a small tree in its native habitat. The *Bombax ellipticum* and *Fockea edulis* also experience periods of dormancy. These are both very old (hence expensive) plants; don't think you can find them at this size at your local nursery. If you do manage to find a large one for sale, you can be sure that someone must have grown them for many years, and despite this long-term attachment, finally decided to or was forced to part with them. Or start small and grow old with them yourself. I've had several starter caudiciforms for 25 years, and they are just now getting impressive.

Adenium obesum, the 'Desert Rose'

Adeniums are some of the most beautiful and uniquely shaped of the caudiciforms. However, like many plants from the equatorial regions of Africa, they don't like cold weather. In California they need to come inside or into a greenhouse throughout the wet winter months. They go leafless and shouldn't be watered until it warms up; cold and dry is usually okay, cold and wet can kill them. For this reason they are not good landscape plants, unless you live in south Florida, Hawaii, or some other tropical location. They are among a handful of "special needs" succulents, but are worth the effort if you're not a lazy succulentist (like many of us).

A *Dioscorea mexicana* dressed up as a sea turtle. Notice the vine growing out of the backside; it winds over 10 feet from the base. Ceramic sea turtle by Jolee Pink.

Dioscorea elephantipes

The "sea onion", *Bowiea volubilis*

The Australian "bottle tree", *Brachychiton rupestris*

The "pregnant onion", *Ornithogalum caudatum*

Kingdom: Plantae

Phylum: Magnoliophyta

Class: Liliopsida

Order: Bromiliales

Family: Bromeliaceae

Genus: Dyckia

Species: brevifolia

Hechtia 'Aztec Gold'

Interlude: *Terrestrial Bromeliads*

Terrestrial Bromeliads

When is a succulent not a succulent? When it looks like one and behaves like one, but technically doesn't fit the description of a "juice-filled" water-storing plant. In general, bromeliads are more tropical, sometimes arboreal plants that rely on rainfall and high humidity in their jungle homes. However, in parts of Central and South America where rainforest gives way to dry and desert-like conditions, the native bromeliads evolved along with the climate and now live happily alongside their distant cactus cousins. Some even have developed thicker and juicier leaves, so you might say they are in the process of becoming succulent. And most importantly, we succulentophiles have decided we like them, both for their appearance and their low-water, low-effort nature.

Dyckias, hechtias, puyas, deuterochonias, and a few other more obscure members of the Bromeliad family cross over somewhat into the succulent realm. They have a traditional root system that defines them as more terrestrial than their arboreal cousins, whose roots serve more as holdfasts. Care and feeding are virtually identical to most succulents. Most of these plants are heavily armed with recurved spines, making vegetative propagation a challenge. Many have intriguing flowers on long stalks, particularly the dyckias. *Puya alpestris* is famous for its amazing metallic sapphire-blue flowers. These terrestrial bromeliads are becoming increasingly popular among collectors and are available on the Internet, but are still elusive in most nurseries. Identification can be an issue as well. I've been in the nursery business a long time and still have trouble telling the difference between a dyckia and a hechtia. Something about where the flower stalk emerges. In general, dyckias have more showy flowers, usually orange. The hechtias are usually larger, but all the hybridizing is making it even more difficult to identify your plant.

The famous metallic teal flowers of *Puya alpestris*, also known as the 'Sapphire Tower'.

I sent a photo of this plant to my friend, Wes, for identification—as he is as close to an expert on these plants as anyone I know. His response was, "Oh, yeah, that one". It may be a form of *Hechtia glomerata*, but I now prefer to call it *Hechtia Oh-yeah-that-one*.

ABOVE *Deuterochonia brevifolia*, a succulent blob in the process of claiming its corner of the shadehouse. This grower's flat of *Deuterochonia brevifolia* has laid claim to the bench it has occupied for close to twenty years. It's not like there's a shut-off switch for a happily growing plant.

FAR LEFT An unidentified hechtia shows its relationship to the pineapple, perhaps the most famous of the Bromeliad family. These arid varieties don't fruit like the pineapple, but can provide impressive flower stalks. As with most succulents, the appeal is much more in the foliage than the brief flowers.

AT LEFT *Hechtia rosea* flower

Dyckia 'Bill Palen'

Hechtia 'Aztec Sun'

Dyckia 'Burgundy hybrid'

An unidentified hybrid dyckia

A seedling version of *Dyckia* 'Arizona Brittle Star'

Hechtia 'Baker's Beauty'

Dyckia texensis

Dyckia 'Precious Metal'

The rare *Dyckia delicata*. Photo by Constantino Galardi.

CHAPTER 10: Succulent Giants

Is there any other botanically-related group where you can keep a collection of thimble-pot sized miniatures on your windowsill, or plant one of its relatives in the yard and have it grow over your house? Here are some succulents that tend towards gigantism.

Succulent collectors can have a balcony of little container creatures, or a yard full of small to tree-size succulents in the ground. Drive through some of the older neighborhoods in California, or visit a few botanical gardens, and you will find some bizarre giants that sometimes seem like a cartoon landscape come to life. Dragon Trees and tree aloes look like something from a Dr. Seuss book, and huge euphorbias and cacti provide a backdrop of colossal columnars. You may have a little plant in a pot right now that yearns to get its feet in the ground and realize its true potential. Check out what a little "elephant foot" beaucarnea can grow into on the next page.

When I consult with customers (at my nursery) on planning a succulent garden, I tell them to think big first, to picture their 'anchor' plants that will grow to a size that will frame their garden. Quite a few will gravitate towards a 5-foot tall, green *Euphorbia ingens*, which will give them the vertical element they may want. Then I tell them to look up. I have a handful of *E. ingens* and *E. ammak* that were 5-foot plants in plastic pots that I set down in places when I opened my nursery 20 years ago. The pots have long-since disappeared, and I now have 25 to 30 foot succulent trees in their place. Be aware of how big your plants will get and how long it takes them to get there.

On these pages are Dragon Trees, *Dracaena draco*, one of the landmark succulents in California landscapes. Native to the Canary Islands, it grows slowly but easily from seed (cuttings are difficult to establish), and a plant as big as the one above is at least 50 years old. Some branch low to the ground, others remain single heads and grow 20 feet high before the first flowering event forces it to multi-head. Once you know what a draco is, it's hard to drive around and not notice these island giants living among us. It is not the true "dragon tree" of antiquity, which is the source of valuable resins; that is likely the related *Dracaena cinnabari* from Socotra in the Indian Ocean. It has a similar profile with stiffer leaves and is rarely seen in cultivation, because it seems to be more cold-sensitive. Habitat photos of Socotra show some of Earth's most amazing plants.

Above is one of the most recognizable succulents, *Beaucarnea recurvata*, also known as elephant's foot, bottle palm or ponytail palm. Most people will recognize it as the fat-bellied potted plant with a baseball-to-beach ball sized base (inset photo). If you live in a climate where you can plant it in the ground, you will get a Volkswagen beetle-sized caudex, essentially becoming a fat tree in your landscape. Large flower clusters cause branching with age. The tree shown above must be at least 50 years old, but at one time it was a little potted houseplant that finally got liberated to be all it can be.

This photo was sent to me by the botanic department of the San Diego Zoo to get an estimate on the value of the specimen. It is almost "whatever someone is willing to pay" for a one-of-a-kind specimen, certainly in the thousands of dollars, likely tens of thousands. But you'd have to consider the several thousand dollars more it would cost to move it. I still don't know if it has moved 'uptown' or not.

The author's bias may be showing a bit here, as there are several nice large branching yuccas that provide a similar succulent-tree look. But in my opinion, tree aloes have a more graceful and artistic shape and silhouette.

Here are three of the four major branching tree aloes. **ABOVE** is *Aloe barberae* (bainesii), a tree that may have inspired author Theodore Geisel, a San Diego resident who wrote children's books under the pen name "Dr. Seuss." **ABOVE LEFT** is *Aloe dichotoma*, a slower and more difficult grower. **AT LEFT** is *Aloe* 'Hercules', a cross between the two, and a vigorous grower. The fourth is *Aloe tongaensis*, seen in the aloes chapter.

Euphorbia ammak variegated

Fouquieria columnaris

Alluaudia procera, also known as the Madagascar ocotillo. In an example of convergent evolution, it has adopted a growth habit that mirrors that of the ocotillo (*Fouquieria splendens*) of the American and Mexican deserts. Although the ocotillo has more impressive red flowers, the alluaudia has a fuller and greener structure, and is better suited for Mediterranean climates. They are deciduous during the cold winter months.

Top images taken at the San Diego Zoo Safari Park.

There are a number of striking columnar succulents. The classic is the "saguaro", *Carnegiea gigantea* (see a photo of the mutated form on page 176). It is much happier in the true deserts than in Mediterranean climates. There are other cacti, and especially several large euphorbias, that offer the columnar statement that announces a dry garden. On the opposite page is the variegated form of *Euphorbia ammak*, which will grow to over 30 feet high like its more common green cousin, *Euphorbia ingens*. It is easily grown from cuttings, and the variegated yellow really makes it stand out against most backgrounds. Another large cactus-shaped tree euphorbia is *Euphorbia abyssinica*. Above is a large "boojum", *Fouquieria columnaris*, next to the "Baja saguaro" *Pachycereus pringlei*, also known as the "Cardon". These are the landmark sentinels of the boulder fields of the Catavina Desert in central Baja California.

Avonia quinaria (formerly *Anacampseros alstonii*) is a highly sought after dwarf caudiciform, with pink or white flowers in spring and summer (as shown in the photo at right).

Interlude: *Odds and Odders*

We have reached the point in this book where the broad categories of succulents have been touched upon, and now all the strange sub-groups deserve a brief sampling. And by "all" I really mean "some", as there are too many obscurities in the succulent realm to get proper treatment here. Some groups are large enough to merit their own page, while others have a token individual that will have to suffice—but be aware that there are many related plants you can find. I hope this interlude provides a jumping-off point for your own research.

PACHYPODIUMS

Pachypodiums are succulent members of the periwinkle family, which includes plumerias. Indeed, these mostly Madagascan plants often bloom with large clusters of plumeria-type flowers. Most drop their leaves in winter, and some are too sensitive to live outside the greenhouse. The most resilient species seem to be *Pachypodium lamerei, P. geayi, P. saundersii,* and *P. namaquanum.*

ABOVE *Pachypodium lamerei,* the Madagascar palm
TOP RIGHT Flowers of the *Pachypodium namaquanum*
AT RIGHT *Pachypodium namaquanum*

STAPELIADS

Stapeliads are clustering succulents that exhibit unusual star-fish-shaped and foul-smelling flowers; hence the occasional moniker of "carrion flower," because the dead-meat fragrance attracts flies instead of bees as pollinators. Both the plant and especially the flower give it an otherworldly feel, and make the plants nice additions to the undersea succulent garden. Related plants, with similar flowers, are carallumas, huernias, and hoodias. Don't let the bad smelling flowers put you off. Outdoors, you have to get very close to get a whiff. However, if one blooms in a sunny, unventilated room, well, you won't leave it unventilated long.

TOP LEFT *Stapelia hirsuta*
MIDDLE LEFT *Stapelia variegata*
BOTTOM LEFT *Stapelia grandiflora*
ABOVE *Stapelia gigantea*

COTYLEDONS

Cotyledons are generally powdery white to gray. These fleshy bushes are very drought-tolerant, and usually have nice bell-shaped flowers on tall stalks. They are not always stunning plants in and of themselves, but make good background companion plants. The 'Silver-dollar jade', although not a cotyledon, blends nicely into this subdued plant palette. Cotyledons are very xerophytic, and many are of the "vacant-lot" category of neglect-tolerance in Southern California. The most common, *Cotyledon orbiculata*, is variable in appearance and form.

ABOVE *Cotyledon orbiculata*
TOP RIGHT *Cotyledon* 'Lady Fingers'
MIDDLE RIGHT *Cotyledon orbiculata* 'High White'
BOTTOM RIGHT *Cotyledon orbiculata* 'Clam Shells'

TYLECODONS

Tylecodons are cousins of the cotyledons, both offshoots of crassulaceae. In fact, their names are anagrams. Tylecodons are typically summer-dormant; leaves fall in late spring as the plant flowers, with long stalks that remain until the new succulent leaves return in the fall. Most Tylecodons have fat tree-like bodies that make them excellent candidates for bonsai. By far the most common, both in African habitat and in cultivation, is *Tylecodon paniculata*, a fat-bodied caudiciform. Slightly less common is *Tylecodon wallichii*, a knobby-bodied mini-tree. There are many more that fall into the rarified realm of collectibles. If you find a *Tylecodon pearsonii* or *T. reticulatus*, you know its owner is a serious succulentist.

TOP LEFT *Tylecodon wallichii*
TOP MIDDLE *Tylecodon hybrid*
TOP RIGHT *Tylecodon paniculata*
BOTTOM LEFT *Tylecodon hybrid*

SANSEVIERIAS

Most sansevierias are thought of as houseplants, but many are succulent and will thrive in partial to full sun. Some of the cylindrical forms make dramatic statements in succulent gardens.

ABOVE *Sansevieria trifasciata*
TOP RIGHT *Sansevieria haemanthifolia*
AT RIGHT Here is the plant that fooled the plant guy. It looked like a slightly bluer form of *Sansevieria desertii*. It was in a heavy terra cotta pot, looked and felt like the real thing, just what I thought was a superior blue form. Then I broke a piece and it was Styrofoam inside. Manufacturers are getting good at fake plants. This one lives in a dark place indoors now (because it can).

SEMPERVIVUMS

Sempervivum ("live forever" in Latin) are clumping succulents that mostly originate in cold-weather portions of central Europe. They are often featured as rock plants in alpine rock gardens, and are one of many plants referred to as "hens and chicks" due to their prolific nature. Most sempervivums would prefer cooler climates than the other succulents featured in this book; however, there are a few that seem to tolerate Mediterranean conditions, including those featured on this page.

TOP LEFT *Sempervivum arachnoideum*
MIDDLE LEFT Unidentified sempervivum hybrid
BOTTOM LEFT *Sempervivum* hybrid (crested form)
ABOVE *Sempervivum tectorum*

PORTULACARIAS

These jade-like succulents are also called "elephant bush" or "elephant grass" because they are a pachyderm delicacy. Garden designers use them as an ornamental bush in the large green form, and as groundcovers or hanging bushes in the variegated or minima forms. They are very durable.

ABOVE LEFT *Portulacaria afra*
ABOVE RIGHT *Portulacaria afra variegated*
AT RIGHT *Portulacaria afra v. minima*

FOUQUIERIAS

The most well known of the fouquieria family are the "ocotillos" of the desert southwest, and the "boojum" or "cirio" of Baja California. These two large succulents are recognized sentinel plants of their respective deserts. The ocotillos (*F. splendens*), above right, can grow to over twenty feet in height, with flaming red flowers at the tips during blooming season (inset photo). They will grow along the coast, but are a little happier in inland or desert communities, where they are popular as landscape plants. Two close relatives from Mexico are *F. diguetii* and *F. macdougalii*, similar to *F. splendens* in leaf and flower, but are more bushy and less vertical in habit. All of the aforementioned can drop their leaves at any time, and coastally seem to spend more time leafless than if grown inland. For this reason the 'Madagascan Ocotillo', *Alluadia procera*, might be a better alternative.

Fouquieria columnaris (formerly *Idria columnaris*, a name still used by many old-timers) is a must-have for Baja travelers who have seen them in habitat. It is a very slow grower, but is a plant you can grow old with. I know of a group planting in Escondido of legally collected 1-foot plants in the early 1960s that are now over 30 feet high. The resident said it took at least 30 years to reach such height. They tend to want to shed leaves and go dormant in the summer, and wake up to new leaves and growth after the first rains and into spring. Another highly-sought plant by collectors is *F. purpusii* from central Mexico, which is almost like a miniature boojum with dainty leaves and a greener trunk. Even more elusive is *F. fasciculata*, a caudiciform plant with a bonsai appeal.

ABOVE LEFT The "boojum" of Baja, *Fouquieria columnaris*
ABOVE RIGHT (and inset) *F. splendens* and a closeup of the flower in bloom

ABOVE LEFT *Bulbine caulescens* is an onion-looking succulent from the lily family that blooms yellow in winter and spring. It can get raggedy and needs thinning out. There is an orange-flowered variety called *Bulbine frutescens*, a less vigorous grower.

TOP RIGHT Cycads are not quite succulents, but they have similar requirements and a similar appeal, and blend nicely into succulent gardens. Cycad collecting can be among the most expensive of horticultural pursuits. Some of the blue cycads, such as *Encephalartos horridus* or *Encephalartos lehmannii*, shown top right, can command prices into four digits when their trunks reach bowling-ball size. The high price is usually a result of age; cycads are generally very slow growers. The most common cycad in cultivation is the sago palm, *Cycas revoluta*, which can also take extremes of heat and cold (as low as 15° F).

BOTTOM RIGHT Perhaps the most succulent of orchids are the epidendrums, sometimes referred to as terrestrial or poor-man's orchids. They are not as popular with true orchidists, but we succulent-philes like them for their succulent-like temperament and wide array of colorful and long-lasting flowers. The foliage can get ratty or beat-up at times, so tuck them into a semi-sun location behind other plants where the flowers can pop up and lend background color. There are many colors to choose from.

TOP ROW (from left to right) 'Sea squill', *Urginea maritima*, is a Mediterranean bulb succulent that is very xerophytic (it is supposedly poisonous). It wakes up from summer dormancy with either tall flower stalks (left) or foliage (middle); at far right is *Anacampseros telephiastrum tricolor*.

MIDDLE ROW (left) is *Mangave* 'Bloodspot', an inter-generic cross of *Agave macroacantha* and an agave relative known as *Manfreda maculosa* (right photo). The manfreda makes a nice garden plant, with large flower stalks in spring and summer. However, I was happily surprised a few years back when I learned it could be crossed with an agave to result in the beautiful plant above left. Another mangave cross is the *Mangave* 'Macho Mocha', with more sure to come.

AT LEFT *Calandrinia spectabilis* is favored for its iridescent purple/fuchsia flowers on tall stalks, similar to a purple poppy. The flowers blow around in the wind, rare for succulents. The plant itself is sort of a generic green bush, and unfortunately grows very fast and eventually becomes messy and leggy. The flower production slows down over time as well, so it is one of the few succulents that may require frequent hacking back or starting over on.

CHAPTER 11: Vertical Gardens

You can grow succulents in the ground, or in ceramic pots, or in old cowboy boots or pickup truck beds. Or...you can make a succulent painting in a frame and hang it on your wall. Or make a living wreath for your front door. As long as you can provide them with some light, soil and water, succulents will figure out how to live in almost any orientation.

Vertical, wall-mounted succulent panels have become popular in recent years. Most of these living murals are constructed in a shallow wooden box, planted flat, given time to grow, and then hung. A shallow layer of planter mix is usually covered with a layer of sphagnum moss, then secured in place with hardware cloth or bird netting. Succulents are then inserted as bare-root cuttings, and are sufficiently rooted and ready for vertical hanging within a few months. Maintenance is an issue. It is best to install some sort of drip system, as hitting it with a hose doesn't penetrate very well. As with wreaths, you need a lot of plant material for best effect, and they tend to dry out quickly, so it takes a bit more effort than most of us succulent folk usually want to expend.

Regarding glass terrariums or hanging glass orbs, as nice as an initial succulent planting might look, these are usually not great long-term homes for your plants. They need outdoor sunlight, but may cook through the glass if placed outside. Lack of drainage is also an issue. "Air-plant" tillandsias may be a better solution.

You can be very creative with succulents. Succulent wreaths can be a challenge to build, because you need a lot of plants for proper density, but the results can be fantastic. If it gets proper light and water, and occasional trimming, you can keep it looking like a wreath for a few years. As it gets straggly, just plant the whole thing into a shallow bowl of succulent mix and it will morph into a huge bowl of succulents. No longer a wreath, but perhaps a stock garden for your next wreath-making attempt.

Succulent wreaths are typically available in anticipation of the holiday months. My customers sometimes balk at the relatively high price, so I point out the amount of time and plant material it takes to properly plant one. I have made one once, and that was enough for me. Wholesale nurseries have both the material and available labor to make them, so I leave it to the artisans. If you still want to make one yourself, try to locate some stock plants that you can make numerous cuttings from, such as the mini-leafed compact jade, *Sedum rubrotinctum*, etc. Frames and moss can be found online or via floral shops. Be sure to plant it dense, like the examples shown in this chapter. Or just go buy one!

Design by Barbara Baker

All wreaths on this page by Brad Brown

Cactus Clubs, Nurseries, and Online Resources

If you are a new enthusiast to the world of cacti and succulents, sit in on a meeting of your local cactus and succulent club (there is one, trust me; every region has a club). There likely will be a monthly or yearly show and sale where you can find true inspiration and plants to buy. Often club members are your best source for rare plants, either by purchase, trades, or just shared clippings. You will find others who may share specialized knowledge in certain rare plant groups. I own a succulent nursery and have a general knowledge of many plants, but will gladly surrender the floor to cactus-club members who have the specialized insight into their particular plant passion. You will also be treated to slide shows of expeditions to Africa, Madagascar, or Mexico and see your favorite plants in habitat.

Succulents are enjoying a surge of popularity, and most general nurseries now have expanded succulent offerings. There are also a handful of specialty nurseries where you are more likely to find rarities as well as specialized knowledge and advice. The Internet can connect you to other sources of plants and information. A good starting point is the portal (http://www.cactus-mall.com), which can lead you to online forums and mail-order sources. Plants can survive a few days in a box, so don't hesitate if you can't find what you want locally. Happy hunting!

Botanic Gardens

For true inspiration, seeing rare and mature specimen succulents in person trumps any images you will see in books or online. Most botanic gardens have a dedicated succulent section or regional gardens featuring succulents. Even in cold-winter locations, many public gardens have conservatories where tropicals or Mediterranean plants thrive year-round.

Public gardens worthy of a visit in San Diego county include the Water Conservation Garden at Cuyamaca College; the San Diego Botanic Garden; and the desert gardens at Balboa Park, the San Diego Zoo, and the San Diego Zoo Safari Park in Escondido. Los Angeles has wonderful desert gardens at Cal State Fullerton, UC Irvine, and UCLA.

Plan a full day (or two) to visit two premier botanic wonderlands just miles apart. The Los Angeles County Arboretum in Arcadia has a vast collection of succulents, including a Madagascan garden full of mature specimens. The nearby Huntington Library and Gardens features the world's largest succulent collection, with huge plants over 100 years old! Indeed, you will see some of the largest examples that exist outside of their native habitat. The Huntington, along with the Arboretum, also has extensive collections of palms, cycads, conifers, etc., truly a plant nerd's dream destination. I recommend visiting in the cooler fall-through-spring months.

A grove of Dragon Trees guards the entrance to the desert garden at Balboa Park in San Diego, CA. Botanic gardens provide inspiration for what your little plants may grow into someday.